Planning for Productive Thinking and Learning

Planning for Productive Thinking and Learning

A Book for Teachers

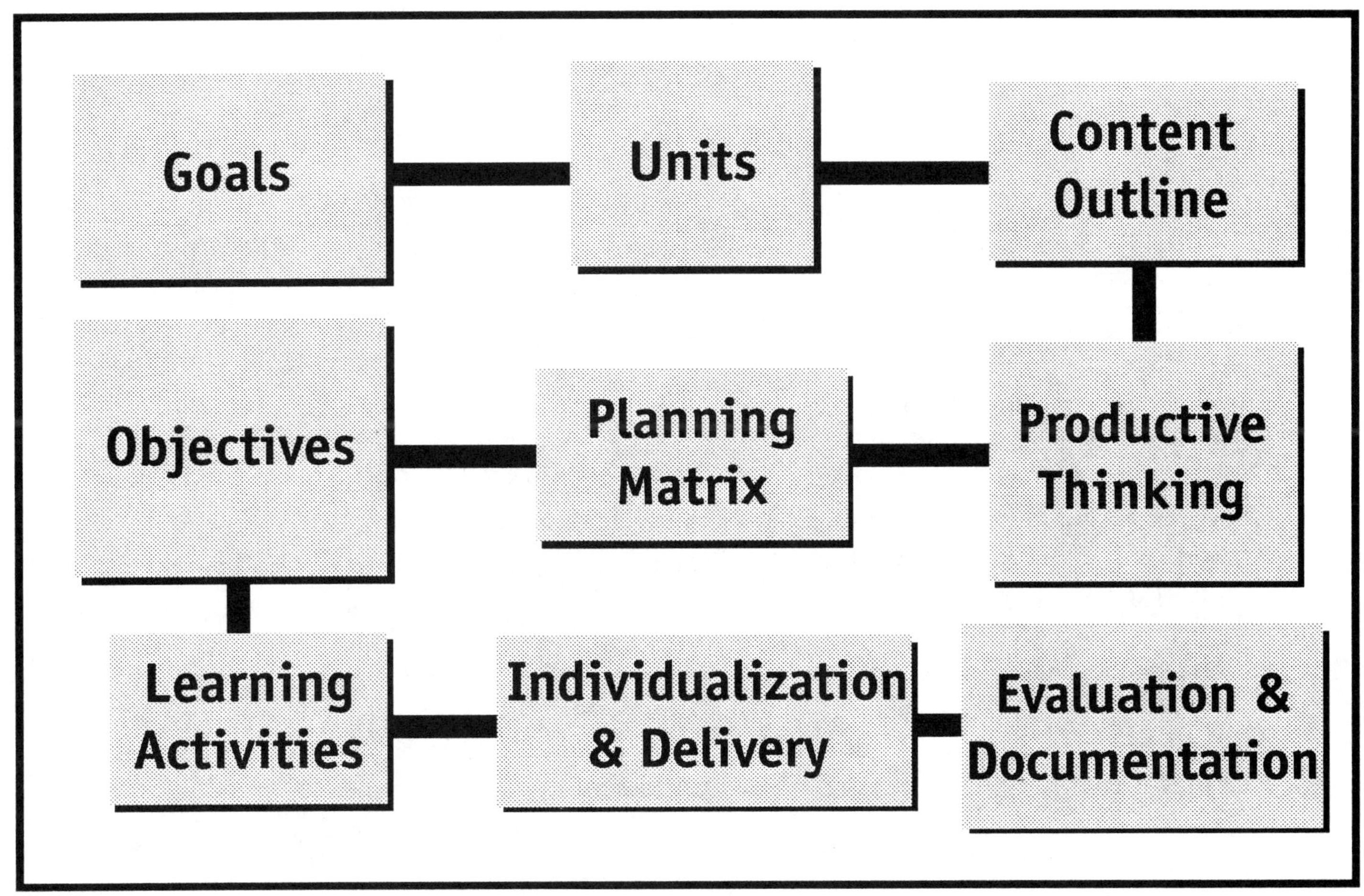

Donald J. Treffinger
John F. Feldhusen

Prufrock Press, Inc. • Waco, TX

Published in cooperation with
Center for Creative Learning, Inc.

Artwork by Barbara Ackerman
Graphic Production by Libby Lindsey

ISBN 1-882664-67-1

PRUFROCK PRESS, INC.
P.O. Box 8813
Waco, TX 76714-8813
Phone: (800) 998-2208
Fax: (800) 240-0333
www.prufrock.com

Contents

Figures

Chapter 1: Goals

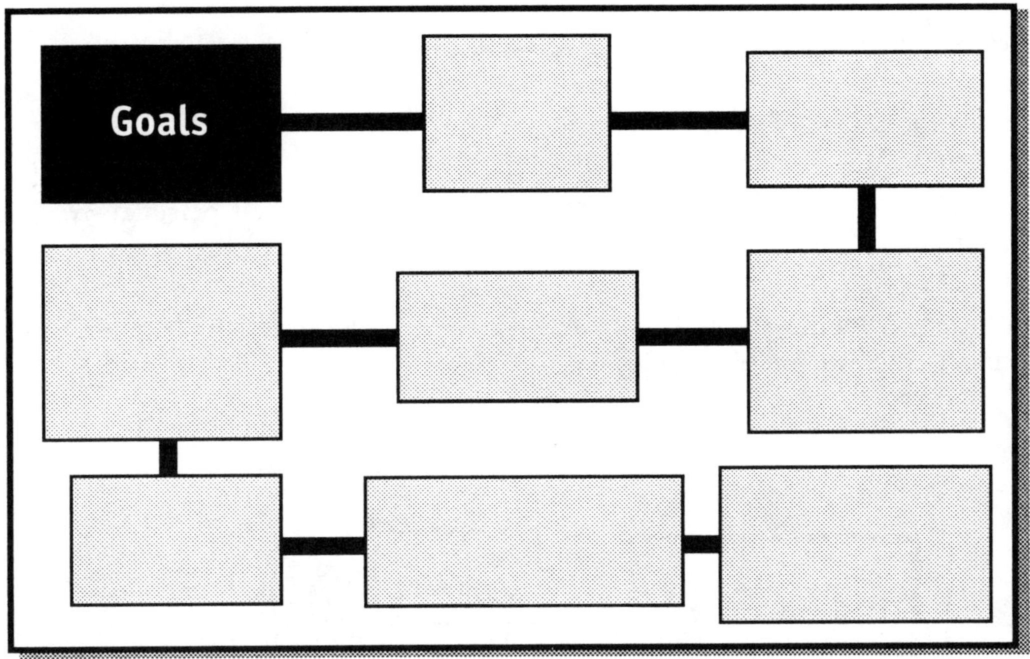

Goals and Objectives for This Book

This book offers you the opportunity to learn and use a number of curriculum and instructional planning tools that contribute to good teaching. As a result of your work in the book, you will be able to:

1. understand the roles of productive thinking, a knowledge base (content, basic information, themes, issues, concepts, principles, problems), student talent strengths, interests, and learning styles in curriculum and instruction;

2. design curriculum and instruction that incorporate a knowledge base, productive thinking, student talents and interests, and learning styles;

3. plan, organize, and conduct instruction using individualized, small-group, and whole-class instruction; and

4. evaluate student progress and achievement using both tests and authentic methods of assessment.

These are very general goals, describing in broad terms what we hope you will be able to do. No single book, module, or training program can provide everything that is important to know about the curriculum and instructional planning. In addition, every educator who reads this book may well be starting at a different point, in terms of present knowledge, current teaching style and activities, or personal and professional goals and interests. You will probably find this book helpful if you hold several beliefs and personal goals, such as:

- a desire to make day-to-day instruction more effective for every student;
- concern for helping your students be productive thinkers;
- desire to help students become aware of, and build upon, their personal strengths, talents, sustained interests, and learning styles; and
- confidence in your own ability to be imaginative in making students' thinking and learning more effective and more enjoyable.

The goals we have described for this book are very important for educators at all levels, from kindergarten through college. Why? First, good planning benefits all students. Well-planned instruction gets students off to a good start, helps the students remember and use what they learn, and is more enjoyable for both students and teachers. Second, good planning helps teachers deal with individual differences among students more effectively. Following the steps in this book will help you in dealing with the characteristics and needs of individual students. Third, systematic planning will improve your ability to promote productive thinking. The process in this book emphasizes specific ways to use thinking processes that might otherwise be overlooked.

Specific Objectives

In order to achieve these goals, this book has 17 specific objectives. By reading the book, working through the suggested exercises, and applying these methods in your own setting, you should be able to:

1. describe four basic elements of a unit of instruction;
2. explain the curriculum guideposts and the instructional guideposts;
3. describe a content outline for a unit of instruction, in a curriculum area that is relevant to your own responsibilities and interests;
4. identify themes, issues, concepts, principles, and problems that should be incorporated into the content of a unit and develop a concept web;
5. plan an original content outline for a unit of instruction, then compare and evaluate the content outline in relation to curriculum standards for that domain or content area;
6. give examples of the six dimensions of the framework for productive thinking and examples of specific methods and tools that can be used to stimulate productive thinking among students;
7. use the productive thinking framework to identify and select specific productive thinking skills and tools to use in the unit of instruction;
8. combine your content outline with the productive thinking skills and tools and apply the curriculum guideposts to create an instructional planning matrix for a unit;
9. define, explain, and give examples of objectives and learning activities and distinguish between them and explain their interrelationship;
10. define and prepare examples of appropriate and reasonable instructional objectives involving several productive thinking skills and tools for each topic in the matrix;
11. develop and use appropriate and varied learning activities, following the recommendations in the instructional guideposts, for a curriculum or instructional unit;
12. know and explain strategies for considering content or tasks, learning styles, and the students' prior experiences when planning curriculum or instructional units;
13. know, explain, and give examples of metacognitive skills and their importance in developing productive thinking;

14. describe nine dimensions of a classroom context that encourages productive thinking and explain ways to establish and maintain such an environment;

15. know many instructional resources and materials appropriate for students' characteristics (e.g., age, achievement levels, styles, talent strengths, special interests) and for the subject matter or content domain; know and apply criteria for reviewing and evaluating curriculum and instructional materials;

16. create an original plan for organizing and presenting the unit as an efficient, enjoyable learning experience with individualized instructional opportunities that take into account students' varied talents, interests, and styles; and

17. plan evaluation procedures for testing and documenting student achievement of the unit's objectives and related concepts or principles using appropriate methods and resources (including tests, rating scales, student products, performance tasks, portfolios, or other forms of authentic assessment appropriate for the task and process skills in the unit).

An Important Note About Time and Pace

Planning a new unit of instruction is not something that can be started and finished overnight. The planning process in this book is intended to be a long-term process that you can work on gradually over an extended period of time. We hope you will use it to begin working on an individualized unit that will continue to grow as you locate or create new ideas, activities, or resources ... over a period of days, weeks, or even months. We hope also that you will develop several units that can be synthesized in a careful plan or scope and sequence for effective instruction throughout a semester or a school year.

Chapter 2: Units

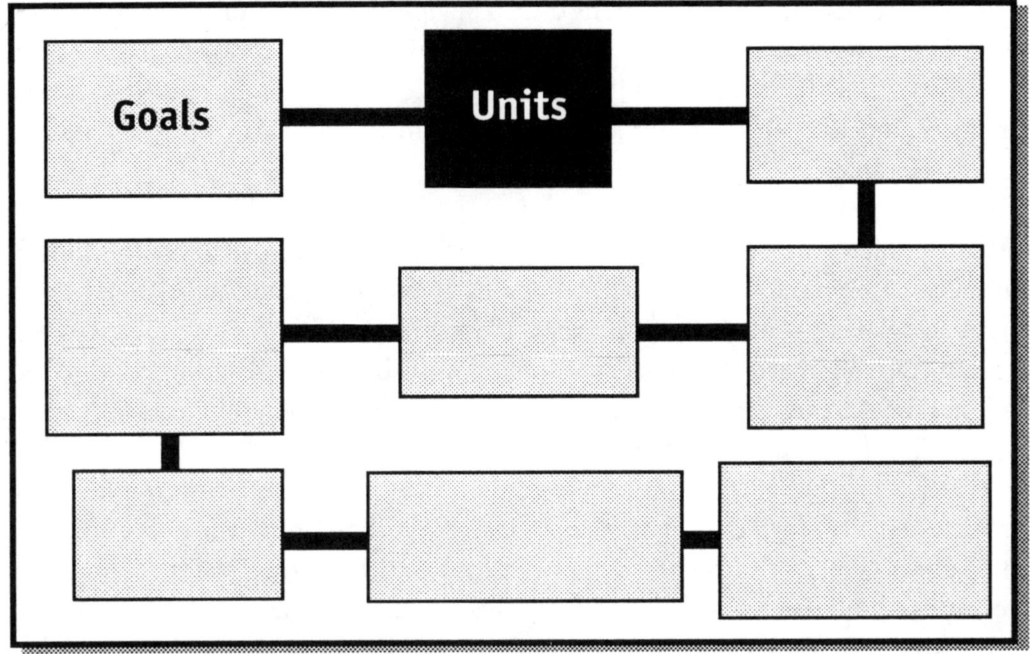

Instructional planning is an important concern at all levels of the educational process, in all subject areas, and with all age levels—from the early primary grades through the highest levels of graduate instruction. Many people share the responsibility for decisions about the content, organization, and delivery of effective instructional units:

- textbook writers;
- state education agency specialists;
- curriculum designers or coordinators;
- curriculum writing committees;
- team leaders;
- curriculum specialists within a school;
- gifted/talented teachers;
- gifted program coordinators;

- special education teachers; and

- classroom teachers.

Whether you are planning a single period of instruction for one or more students, making a grade-level plan for several classes, developing a special program, creating a curriculum for an entire school district, or developing a syllabus or course of study for an even broader audience, you will be concerned with the question: "How might this material be designed and organized as effectively as possible?"

Most educators have been planning instruction ever since their student teaching days. At some point in your preparation for teaching, or early in your teaching career, you probably learned (or, at least, were told) how to prepare a "lesson plan." For some teachers, this was a major topic in a teaching methods course; for others, it probably involved meeting requirements for a weekly plan book; and, for yet others, it may have been learned by trial and error as a way of surviving the first few years of teaching.

If your experience is like that of many educators with whom we've worked, your early instruction or experience in "lesson planning" involved outlining the content to be covered, the activities to be used, questions to be asked, motivational strategies, evaluation methods, and probably some kind of culminating activity or project. And, it is quite likely that you learned, as many educators have been taught for years, that a "unit" is simply a collection of lesson plans organized around a certain content topic or theme.

As you gained experience in teaching, however, many of you probably stopped doing the elaborate written planning that you learned in those early days or years for various reasons:

- Perhaps, as the content became more familiar, you began to spend less time writing out lesson plans and devoted more time to looking for or creating new resources or activities.

- Perhaps you became concerned that excessive focus on details and very specific written plans might stifle curiosity and spontaneity (the students' and your own) in your classes.

- Perhaps, with the growing number of things that must be dealt with in the school's curriculum each year, you found it difficult to meet the ever-increasing demands on your time and energy for written unit plans.

It may be helpful, then, for us to guide you in taking a fresh, new look at the whole subject of units and instructional planning. The approach that will be presented in this book is realistic (if challenging), practical, and useful in a number of different ways. Examples of some of the ways various groups of educators have already used this approach are:

- individual teachers have planned specific units for students in their classes;

- groups of teachers have changed existing curricula and developed new curricula for specific grade levels, subject areas, and special programs;

- instructors and trainers have organized and conducted courses, workshops, and seminars on instructional planning and in professional development programs on individualization and curriculum design; and

- curriculum specialists and authors have designed a variety of instructional materials on interrelated themes or topics.

What is a Unit of Instruction?

As we will use the phrase in this book, a "unit of instruction" has four general characteristics:

1. **Content Objectives.** Identify the major learning outcomes for the students.

2. **Productive Thinking Skills.** Identify several thinking skills for the students to learn and use, going beyond recognition and recall.

3. **Learning Activities.** Identify one or more appropriate learning activities for the students for each objective.

4. **Evaluation Procedures.** Specify the procedures that will be used to evaluate student progress for each objective.

As we define it, then, a "unit of instruction" is a *systematic plan that is created to guide your teaching with individual students and classroom groups.* It is an orderly way to "map out" many and varied instructional options that you can use with your students. It is not a lesson plan in the familiar, traditional sense, since it does not prescribe in a step-by-step sequence how you will actually teach the content of the unit. Instead, it may be helpful to think of a unit as your "master plan" to guide you in your decision making as you organize the material or work with your students.

A unit of instruction helps you plan more effectively—in any content area—for a wide range of alternative activities or experiences from which you can draw for your students. The unit of instruction can be used in many ways. For example, units can be used as a way of:

- planning day-to-day teaching strategies for individual students or groups;
- organizing self-directed learning packages or modules;
- organizing learning stations, learning centers, or interest centers;
- planning for instruction using contracts or learning agreements;
- designing a content plan for a complete course or program;
- creating plans for books, media programs, or computer programs through which instructional content is presented; and
- organizing record keeping for students' progress and accomplishments.

A unit of instruction is also based on several important principles that we refer to as *curriculum guideposts* and *instructional guideposts* (Treffinger & Feldhusen, 1996). These two sets of principles are presented on the following pages. We will discuss their applications in unit planning in greater detail later in the book. Throughout the book, we will also include examples of each of the important parts of the unit of instruction. In addition, we will guide you in designing and organizing your own units of instruction.

Curriculum Guideposts for Productive Thinking

1. Help students learn basic knowledge (information, concepts, themes, and so forth) at the highest or most advanced levels appropriate for their age, achievement levels, and grade.

2. Teach methods for productive thinking.

3. Teach methods for research, inquiry, question asking, and independent study.

4. Teach methods for self-direction and control of learning (metacognition).

5. Provide learning experiences in curricular areas related to student talents, interests, and learning styles.

6. Provide curriculum experiences (in or out of school) that relate to real life experiences, careers, and applications.

7. Teach methods for, and encourage self-appraisal and reflection in, learning.

8. Include issues, controversies, and problems that are unresolved, ambiguous, and indefinite.

9. Organize learning experiences so that students can recognize, create, and work with themes and patterns.

Instructional Guideposts for Productive Thinking

1. Offer a variety of instructional methods that allow different learning styles to be recognized and used.

2. Encourage students to use self-regulated, self-monitoring, and self-directed learning.

3. Set high standards and expectations for student performance, achievement, and products.

4. Provide for active involvement of students in learning situations.

5. Provide time and opportunities for student reflection about, and discussion of, their learning experiences.

6. Use a variety of arrangements (whole class, small groups, and individual) and activities as appropriate for the objectives.

7. Provide question-asking, research, inquiry, and independent study opportunities for students.

8. Develop learning environments that provide both order and freedom and encourage productive thinking and learning.

9. Engage and challenge students at all levels of productive thinking and at the highest and most complex levels as their readiness emerges.

10. Strive for student passion and engagement in the learning process.

11. Provide opportunities for intrinsic and extrinsic motivation and for cooperation, collaboration, and appropriate competition in learning.

12. Use a variety of media, resources, and technologies in teaching.

13. Provide opportunities for students to use what they have previously learned in new situations.

14. Provide leadership to respond to students' varied talents and interests.

Suggested Activities

1. As a self-check, review the content of this chapter by listing and describing the four important components of a unit of instruction.

2. Write your own summary, describing how our view of a "unit" differs most importantly from more traditional ways of describing "units of study."

3. Review the Curriculum Guideposts and the Instructional Guideposts and make notes on how, and to what extent, you are already applying them in your teaching.

4. Make a list of several topics or themes of particular interest to you that might be developed into units of instruction.

5. What parts of an effective unit of instruction are the easiest for you to prepare? Why?

6. What parts of an effective unit may be the most difficult for you to prepare? Why?

CHAPTER 3: CONTENT OUTLINE

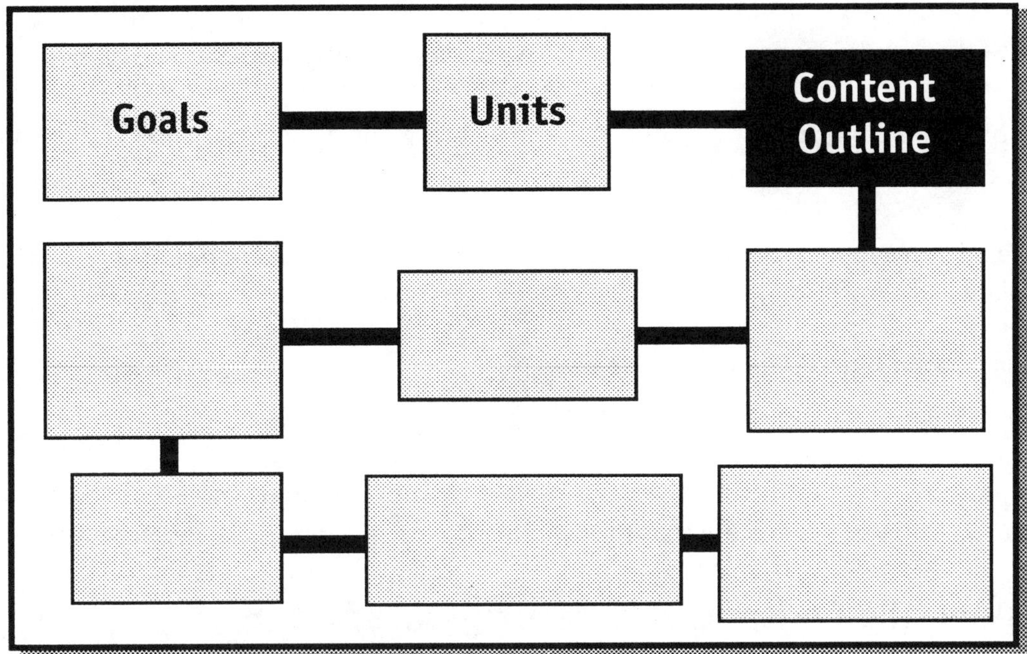

Once you decide to develop a new unit of instruction or revise existing materials that you've used in the past, one of your first steps is to begin thinking about the content of that unit. Content refers to the general topics or subject matter that will be included in the plan (which we commonly describe as "what will be covered" in the unit).

Many teachers begin planning a new unit by asking the question, "What are some of the questions or topics my students should have an opportunity to study during this unit?" This is a question that concerns the content goals for your unit.

At the beginning of your planning, then, you are probably not concerned with spelling out in detail all the facts, information, or principles that your students will learn. Rather, you are concerned with identifying the broad topics, concepts, themes, or issues that you want to include. These might be organized or listed in an outline format. This chapter will help you plan a content outline for a unit of instruction.

Why Make a Written Outline?

Many teachers rely on doing instructional planning "in their heads" when they are developing or revising a unit. Still, there are several good reasons for writing out your plans. After you have actually completed all the steps in this book for a unit of your own, we hope these reasons will be clear to you. But, it may also be valuable to begin by sharing with you some of our reasons for beginning with a written content outline.

The written outline will be:

1. valuable to you later in the planning process (such as planning learning outcomes and activities);

2. helpful in your efforts to coordinate the various parts of your unit;

3. helpful in guiding your decisions about evaluating the appropriateness and relevance of the content of the unit;

4. valuable in your efforts to be certain that important topics and issues have not been overlooked;

5. useful in your future efforts to review and classify new resources that might be included in revisions or expansions of the unit; and

6. helpful in your consideration of ways to use the unit material with students, to organize your presentation of the unit, and to document students' products and accomplishments.

Where Does the Content Outline Originate?

When you begin thinking about the content of your unit, there are at least three important sources of information:

- your own imagination, ideas, and memory bank;
- your students' characteristics, needs, and interests; and
- professional resources and published curriculum materials.

We will consider each of these sources briefly now and in greater detail in the rest of this chapter and in the next chapter.

Your own memory bank. When you are planning a new unit of instruction, you should use your own memory—a rich storehouse of useful ideas and activities. Through your experience in teaching and your knowledge of the material in the unit, you have learned quite a bit about which subjects are most interesting to you and which will also arouse the interest, "spark" the curiosity, and hold the attention of your students. Thus, at the beginning of your thinking about the unit, take some time to list the general themes, topics, or issues that you will want to include.

Your students' characteristics, needs, and interests. The content of a unit should also take into account the background, experience, needs, interests, and preferences of your students. The special abilities and talents of your students and the unique learning characteristics of students who differ in ability should also be considered when you are planning your unit.

Professional resources and published curriculum materials. One of the traditional sources of content for a unit of instruction is published material. Most teachers use books, films, filmstrips, videos, computer programs, and other learning aids in the course of developing a unit of instruction in their classroom. You are probably always on the lookout for new ideas or materials that will be useful in your teaching.

When you are developing your content outline, you should take advantage of any ideas that you can obtain from such sources as:

- textbooks;
- supplementary or enrichment books;
- publications by public, private, or service organizations;
- audio-visual materials;
- teachers' guides;
- programmed instructional materials;
- learning modules or mini-courses;
- CD-ROM programs or other interactive multimedia resources;
- published learning center resources;
- material from general magazines for teachers (such as *Instructor* or *Today's Education*) or from more specialized publications (such as *The Reading Teacher* or *The Mathematics Teacher*);
- local or state (or provincial) curriculum guides or resource books;
- publications of professional organizations;
- curriculum resources from special or experimental projects;
- ideas and materials from workshops or conferences;
- topics suggested by television, newspapers, or news magazines;
- the Internet and the World Wide Web;

- magazines written for members of a general audience who are interested in a specific area (such as *National Wildlife*, *National Geographic*, or *Ranger Rick*); or
- colleagues or other people in your area who are willing to share their talents, experiences, and resources.

How to Prepare the Content Outline

There are at least two useful ways to prepare a content outline, which you can use separately or together. These are *concept webbing* and *question clustering*. Each of these is easy to learn and use.

Concept Webbing. Concept webbing is one creative and often productive way of generating and pulling together the general topics for a content outline for your unit of instruction. You can begin by selecting a broad concept, or a "chunk" of content or information you wish to teach. Broad concepts might include, for example, any of the following: justice, love, change, philosophy, or competition. Sample chunks of content or information might be any of the following: chemical elements, the poetry of Wallace Stevens, the Plains Indians, simultaneous equations, election systems, or Impressionist art.

The webbing process begins, as shown in Figure 3.1, by writing the name of the broad concept or content chunk in an oval in the center of a large space. Then, begin a "brainstorming" or "free association" process, through which you generate and extend the other first-, second-, and third-order circles. What major ideas come to mind, for example, when you think about a unit on the broad concept of family? What subtopics then come to mind for each of those major ideas?

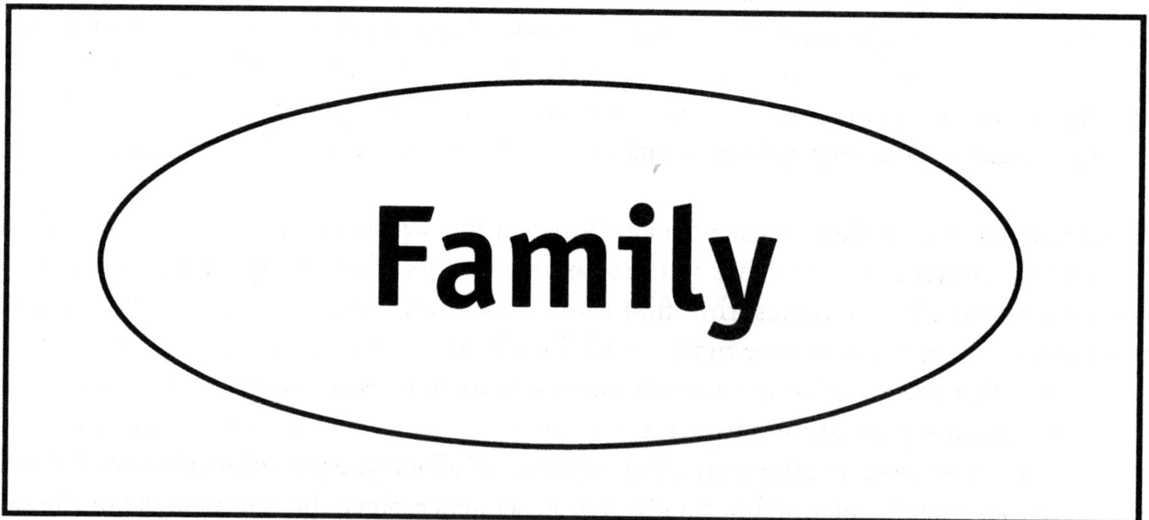

Figure 3.1: The Starting Point for a Concept Web

Figure 3.2 presents a sample of one such web that might be developed for the broad concept, "Family." This web, like most webs for broad concepts, is dynamic, of course, and might be extended a great deal or developed quite differently by various people.

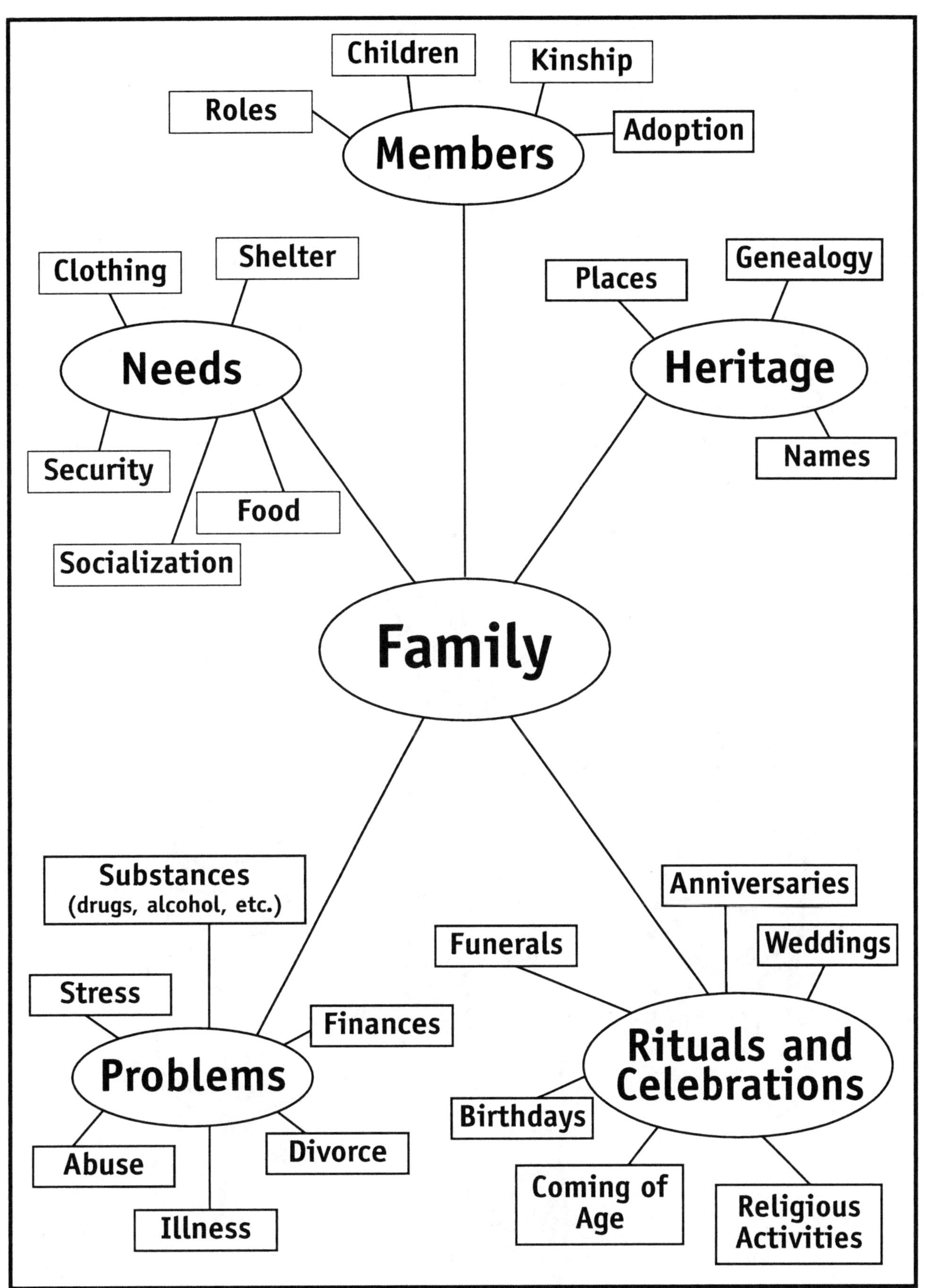

Figure 3.2: Sample Concept Web for "Family"

Question Clustering. This is an open-ended, flexible technique in which you use brainstorming tools to generate possible topics for the content outline, followed by a more critical or convergent clustering or categorizing strategy to draw together the major headings for the content outline.

To use this approach, you begin by gathering all the ideas you can find from many sources. Consider ideas from all of the three source areas described above: your knowledge and imagination, your students' ideas and interests, and published resources to which you have access. Armed with these resources, the first step is to write down every idea that comes to mind. Don't worry about how broad or specific the ideas are and don't try to judge whether each idea that hits you is appropriate or useful. Your goal is to make your list as exhaustive and wide-ranging as possible. It will be much easier to discard the ideas that turn out later not to be useful than it is to try and remember a really unusual idea that wasn't written down. After several hours or days of other work, some unrecorded ideas will be permanently lost. A good suggestion is simply to cover a wall with several large pieces of blank chart paper or newsprint on which to write down the ideas that occur to you.

If you seem to run out of ideas, here are some tips that might be helpful for you to try:

- Put the topic aside for awhile, but carry a small notebook. You are likely to find that, as the incubation process begins to work, you'll think of new ideas when you are away from your charts. Write them down so they can be transferred to the master chart later.

- Try out the topic on someone else by asking, "What questions might you want to ask about [the topic]?" Try this with your students. You will probably discover that they can generate many questions and ideas, even about topics they have not formally studied. (Don't underestimate the extent of ideas students read or hear about in the world.) You may also find that friends and associates from outside school will give you helpful ideas. People who are not experts in the topic may be quite helpful by suggesting unusual questions or original directions that had not been explored. Helpful hint: Do not turn off ideas from others by saying, "No, that won't work," "I don't like that one," or "I've already got that one." Just make a note of all their ideas. You can sort them out later on your own.

When you have generated a good collection of ideas and questions, organize them into several general topic headings. Look for the groupings of questions and ideas that seem to deal with a common topic or subject. Note the ideas that seem to "belong together," and ask, "What do these ideas all have in common?" Use your response to that question to name the clusters. In this step, you are working toward collecting all the questions and ideas (which might be compared in general to an "index," although disorganized; they're very specific entries) and reducing them into a smaller, more compact set of descriptors (which might be better compared to a "table of contents") that will serve as the basis for the content outline. The number of categories you use does not make too much difference. As a rule of thumb, experience suggests that fewer than 4 or 5 categories is likely to indicate categories that are too broad (and will be cumbersome to work with later in the process), whereas more than 9 or 10 may be too many, suggesting that you are still dealing with very specific items (the index again, not the table of contents).

Finally, arrange your shorter list of major topics into a sensible, logical sequence. This will yield a working content outline to use in the subsequent steps of planning the unit. Don't be too concerned with detailed subtopics; the "first level," broad headings, will probably be quite adequate for the next steps. If you have more detailed subheadings, they will help you later to ensure that you haven't omitted anything you really wanted to include. Also, don't worry about whether or not you're absolutely certain the topics are in the "right" sequence. You can always modify the sequence later, if you wish. Furthermore, the sequence in which you plan the unit does not mean that the activities must be planned or taught in that sequence, since the unit plan is not the same as a traditional lesson plan.

An example of a list of ideas and questions generated and then categorized using the question clustering techniques is shown in Figure 3.3. The example deals with the topic, "Colonial America." Many of the ideas and questions in this list were generated by students in the fourth grade, but the same topics might be addressed by students at any age or grade level.

1. Where were the colonies?
2. What work did the men and women do?
3. What kind of government did the colonies have?
4. How did the colonies become independent?
5. What were prices like?
6. What did the boys and girls do for fun?
7. Why were the colonies established?
8. What were their schools like?
9. What were their churches like?
10. How were the colonies different from one another?
11. Who were their leaders?
12. Colonial cities—where did people live?
13. What were their houses and streets like?
14. What did people eat?
15. Where did people get their food?
16. Did they go fishing and hunting?
17. Did colonial people have many friends?
18. What did their flag look like?
19. How many people lived in the colonies?
20. What kinds of farms did they have?
21. Were there more rich people than poor people?
22. How did they earn their living?
23. What kind of candy did they have?
24. What were their different religious beliefs?
25. How were their clothes different from ours?
26. What kinds of music did they have?
27. What was the weather like in the colonies?
28. What kind of houses did they live in?
29. How did they take a bath?
30. Did they have fire and police stations?
31. What kinds of tools did they use?
32. How did the colonists decide where to live?
33. Did they have desks like ours in school?
34. Did they have doorbells?
35. Were children punished in school?
36. Did they have books and libraries?
37. What holidays did they celebrate?
38. Were they very good in sports? Which ones?
39. Did they have parties?
40. Did they have cars and buses?
41. What did children do after school?
42. What kind of money did they use?
43. What did the families do for recreation?
44. How did they communicate with each other, from one colony to another?

1, 7, 10, 12, 19, and 27 have to do with **Identifying the Colonies**

3, 4, 11, 18, and 30 have to do with **History, Government, and Politics**

13, 14, 15, 23, 25, and 28 have to do with **Food, Clothing, and Housing**

29, 31, 32, and 34 have to do with **Home Life**

2, 5, 20, 21, 22, and 42 have to do with **Economics**

8, 33, 35, and 36 have to do with **Schools**

6, 16, 17, 26, 38, 39, and 43 have to do with **Sports, Recreation, and Social Life**

9, 24, and 37 have to do with **Religion and Beliefs**

40 and 44 have to do with **Transportation and Communication**

These questions led to the nine categories.

Figure 3.3: Sample Questions and Clusters for "Colonial America"

A Working Draft of the Content Outline

After you have generated a good list of topics and questions for your unit, using either the concept webbing or question clustering approaches, you will find it quite easy to develop a working draft of the content outline for your unit. As examples, sample content outlines derived from a concept web on "Family" appear in Figure 3.4 and from a "Colonial America" unit in Figure 3.5.

Family

I. Membership
- A. Roles
- B. Kinship
- C. Children
- D. Adoption

II. Needs
- A. Food
- B. Clothing
- C. Socialization
- D. Security
- E. Shelter

III. Problems
- A. Divorce
- B. Stress
- C. Abuse
- D. Illness
- E. Finances
- F. Substances

IV. Rituals and Celebrations
- A. Coming of Age
- B. Weddings
- C. Funerals
- D. Birthdays
- E. Anniversaries
- F. Religious Activities

V. Heritage
- A. Places
- B. Names
- C. Genealogy

Figure 3.4: "Family" Topic Outline

Colonial America

I. Identifying the Colonies

II. History, Government, & Politics

III. Food, Clothing, & Housing

IV. Home Life

V. Economics

VI. Schools

VII. Sports, Recreation, & Social Life

VII. Religion & Beliefs

IX. Transportation & Communication

Figure 3.5: "Colonial America" Topic Outline

Practice Your Own Content Outline

To conclude your work with this chapter and before you continue studying the rest of this book, please spend some time working on the development of a content outline of your own. Choose a topic (a broad concept or a general "chunk" of content) that grows out of the broad scope and sequence of your school's curriculum for science, social studies, language arts, math, or any other content area. Or, select a topic that you think will be valuable for you to work on and interesting both to you and your students. Follow the procedures and steps that were described in this chapter to produce a working draft of your content outline.

Don't expect to complete this in one sitting. The book is intended to be a resource tool with which you work over an extended period of time, not just something to sit down and read. You will learn more and use these ideas more effectively in your own planning if you work through the exercises and activities in each chapter with your own material. (By the way, working together with a partner or a team is a good idea.)

As you begin your practice, ask, "What will be my general topic?" Then, use blank paper or a chart to develop your concept web or to try the question clustering method. Finally, organize your list or web into a working draft of the content outline.

When you have completed your draft, look it over carefully. Check your organization. Would your major headings make good titles for the chapters of a book on the topic? Is the sequence logical and appealing?

Any outline can be changed each time you examine it, of course. Many times, you will find productive ways to modify and expand your unit after you have begun working with a group of students. The content outline is not "carved in granite." It is a flexible tool that you will be able to use as you continue your planning.

You have now overcome the first hurdle. You have identified the content or concept for a unit of instruction that will most likely reach beyond the immediate confines of a specific textbook. You have organized the material in a way that will make the next steps of planning an effective unit of instruction much easier.

CHAPTER 4: PRODUCTIVE THINKING

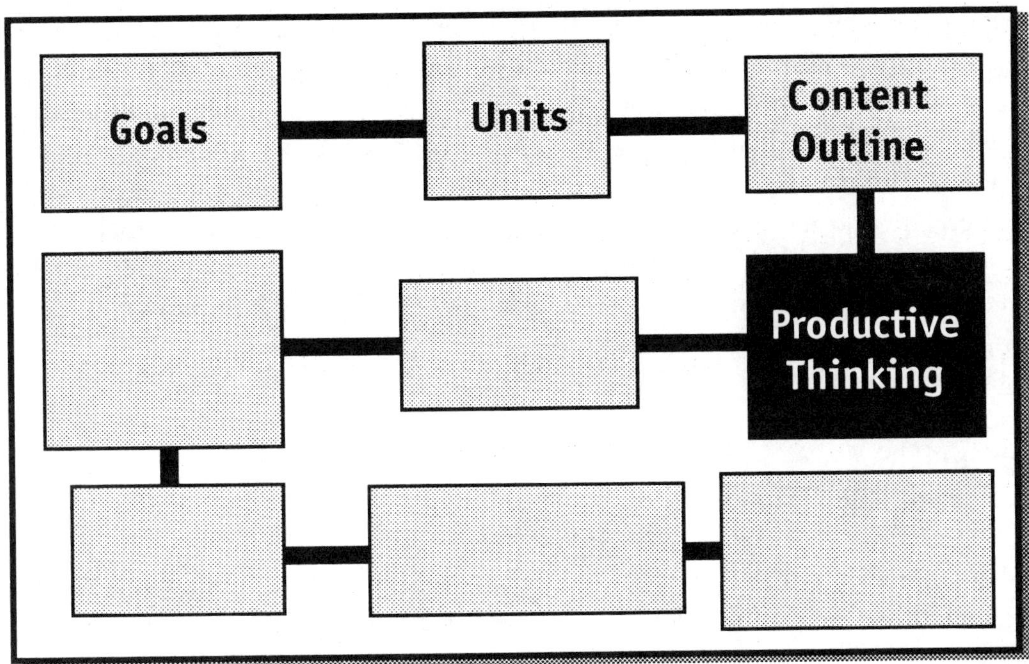

Our focus in the last chapter was on the content of your unit of instruction. Your content outline describes the topics that you expect students to study and learn during the unit. This content is a very important part of any unit.

When you were developing your outline for a content area (such as "Colonial America") or for a concept or theme (such as "The Family") in chapter 3, you may have thought, "These broad topics are rather vague. What exactly will the unit cover? What exactly will students learn about colonial America or about the family?" These are good questions. General decisions about what topics to "cover" are not sufficient for good instructional planning. As a teacher or a curriculum writer or planner, you must also consider more specifically what the instruction will include. For example, in the outline on Colonial America, section VII deals with Sports, Recreation, and Social Life. Should this section cover information about what kinds of sports or recreational activities existed, in which colonies they were found, and what might be unique or unusual about them? Or, should students be able to compare the various recreational and social activities in colonial times with those of other places at that time or with our present day activities? Should we expect that students will be able to explain why different activities might have been preferred in certain colonies or the implications of their social activities for other aspects of colonial life? To be sure, there might be quite a number of specific directions to pursue as the unit plan takes shape. Your expectations might include outcomes for the students to learn and remember, but a number of other, more complex kinds of intellectual outcomes may also be incorporated into your plan.

In addition to content (the *what* of student learning and achievement) we also need to be concerned with students' thinking skills or mental processes (the *how* in learning). Thinking is a very important concern in modern education, since today we have a much richer, fuller understanding than ever before of the powerful computer inside the human head. Treffinger (1988) suggested that good educators have learned that hardware is not very useful until good software is available to put in it. Thinking skills provide that software for the mind. We also know that huge amounts of memory storage are not very helpful unless that memory can be organized and managed well. Similarly, student learning will not be as effective as it could be if we emphasize only remembering information. In people, as in our computers, lots of memory is an asset, but only when there are also more complex or higher level programs or tools to enable that memory to be used effectively. Treffinger, Feldhusen, and Isaksen (1990; 1996) used the term *productive thinking* to refer to four important dimensions of higher level thinking:

- creative thinking;
- critical thinking;
- problem solving; and
- decision making.

They also emphasized that these four dimensions of productive thinking build on a foundation that includes: a rich knowledge base; personal characteristics and styles; metacognitive skills; and an effective climate or environment for learning.

In this chapter, we are focusing on the skills and tools that make up productive thinking, since these will also be important parts of your unit plan. We will help you design a unit of instruction in which all students will have opportunities to learn and apply productive thinking, while also learning appropriate content. This point is important to emphasize because a frequently heard concern among teachers is that they do not have time to deal with thinking skills along with everything else they are responsible for teaching. With increasing pressures to cover more and more content, they say, there just is not sufficient opportunity to deal specifically with thinking skills.

The new constructivist movement in education teaches us that students learn content or subject matter better when they have to think about it, solve problems in it, evaluate or judge it—and then make it a part of their own personal "bank" of ideas and information. Without such constructive activity, students simply memorize content and lose the information soon after the test or the end of the lesson.

Defining Productive Thinking

The subject of thinking skills and tools has become quite fashionable in recent years, with a variety of new journal articles, books, and programs appearing regularly. While a few writers treat this as a really new or innovative concern—a real "discovery" in the North American educational scene—it can easily be demonstrated that systematic concern for teaching thinking has been apparent for more than three decades of educational theory and research. Whether it is a new or old concern, the renewal of interest in thinking skills is an important direction in modern education. Many different models and approaches each call our attention to the importance of thinking that goes beyond memorization and recall.

The variety of different thinking skills models and programs can be confusing to teachers as they attempt to decide which approach will be most suitable or appropriate for their needs and circumstances. A comprehensive listing of every model or published program is not possible in the limited space of this chapter. Many contemporary models and programs have been described by Costa (1991) in a useful anthology entitled *Developing Minds: A Sourcebook for Teaching Thinking.* A number of published resources for teaching productive thinking have also been described and reviewed by Feldhusen and Treffinger (1985) in *Creative Thinking and Problem Solving in Gifted Education* and by Treffinger, Cross, Feldhusen, Isaksen, Remle, and Sortore (1993) in the *Productive Thinking Handbook, Volume I.*

Treffinger, Feldhusen, and Isaksen (1996) presented guidelines and criteria that can be used in selecting or developing resources for teaching productive thinking. These included:

- Is there a sound theoretical foundation?
- Is there effective balance between ease of use and need for training?
- Is the program responsive to individual differences?
- Does the program provide for curriculum relevance and "stretching?" (Can it be applied in curriculum areas? Does it suggest new directions for curriculum development or extension?)
- Is the program based on sound principles of instructional design?
- Does the program guide decisions regarding structure, scope, and sequence?
- Does the program draw methods and techniques effectively from a variety of models?
- Are exercises and activities appropriate for many social and cultural groups?
- Does the program provide for the development of metacognitive skills?
- Are clear, practical examples provided and appropriate applications suggested?
- Does the program respond to student interests and motivations?
- Does the program provide for active involvement and experiential learning?
- Is the program accompanied by (or supported by) appropriate assessment and evaluation resources?
- Is the program supported by research and evaluation evidence of its impact or effectiveness?
- Is the program presented in an attractive and useful format?

Applying these criteria will help you make good decisions about which thinking skills models or methods may be most appropriate and effective in your own teaching. There is no "best" model for everyone to follow to help all students become better thinkers, nor any one program that "fits" the approach to unit design presented in this model. We believe that this approach to unit development for productive thinking can be used harmoniously in conjunction with any of the major approaches to defining and fostering productive thinking. We have created a general framework that can be used effectively to guide teachers as they incorporate productive thinking into a unit of instruction. This framework is presented in Figure 4.1.

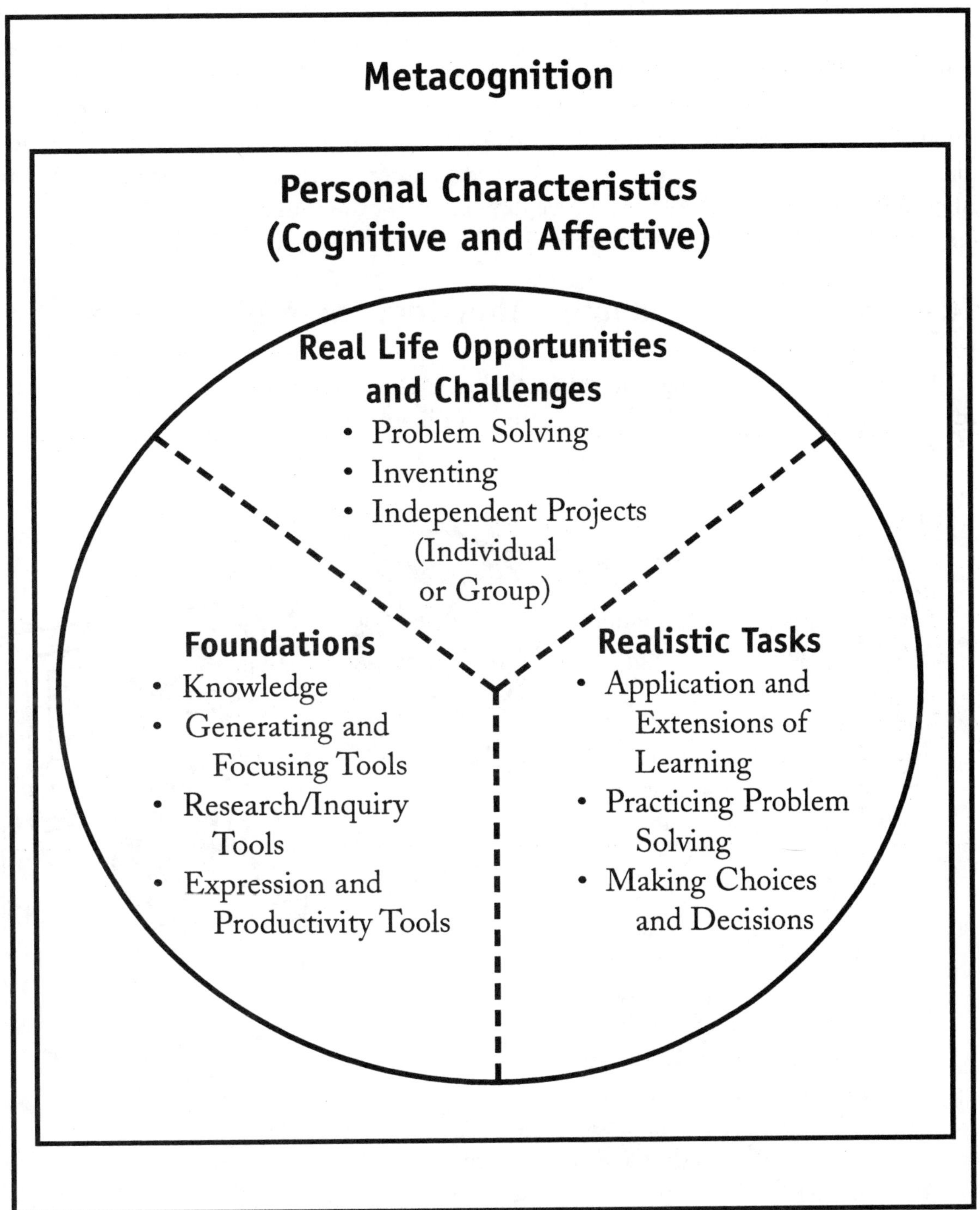

Figure 4.1: A Framework for Productive Thinking Instruction

Examining the Framework

The framework for productive thinking instruction includes three dimensions of skills and tools in the center circle of the figure. These are: foundations, realistic tasks, and real opportunities and challenges. These are arranged in a circular presentation to indicate that they are not hierarchical or sequential. A dashed line is used to separate the three dimensions to emphasize that they are not rigidly separated or isolated categories. We will discuss each of these in greater detail in this chapter, since they will be important elements during the initial stages of unit design or planning. The circle in the figure is surrounded by three concentric squares: personal characteristics, metacognition, and context. These dimensions of the framework are important factors in individualizing and teaching a unit of instruction. They will be discussed in chapter 8.

Skills and Tools for Productive Thinking: Three Dimensions

What are the three important dimensions of productive thinking that should be considered in curriculum development or in designing units of instruction? What specific skills and tools might be included in each of these dimensions?

1. **Foundations.** The foundations dimension includes: knowledge, or rich, well-organized, and readily accessible information; specific tools for generating options or possibilities and for focusing the options that are generated; methodological tools for research or inquiry; and tools and skills for productivity in expressing or presenting the results of one's thinking.

Knowledge. Knowledge involves information, definitions, data, ideas, concepts, beliefs, or theories. Some knowledge comes to us as facts or information to be received or remembered. Knowledge can be as simple as the fact that Miami is a city in the state of Florida or that George Washington was the first President of the United States. We read, see, or hear knowledge as information. More complex thinking draws upon, or builds up, what we know and remember.

Tools for Generating Options. The foundation for productive thinking also involves knowing and using several tools for generating options. An "option" is an idea or possibility that we construct in our mind. There are many times when it is important for people to think of many possibilities, to look at a task or a problem in a different way or from a different point of view, or to be able to come up with new or unusual possibilities. People also have to be able to take a basic idea and expand on it, add details to it, or make it a more complete and interesting possibility. Tools are devices we use deliberately to get a job done. There are "tools for the mind" that help people generate many, varied, or unusual ideas and possibilities. The tools for generating options are often associated with "creative thinking."

Tools for Focusing Options. The word *focus* can help you understand the purpose and importance of these tools. Have you ever taken a picture and discovered later that the camera wasn't focused properly? If so, recall how the image looked: blurry, not clear, perhaps not recognizable at all. The foundation for productive thinking also involves knowing how to use some mental tools that will help students be sure that their ideas are in clear, sharp focus. These tools help people organize, analyze, refine, strengthen, evaluate, judge, or choose among the ideas or possibilities they have generated. These tools are often associated with "critical thinking."

Research and Inquiry Tools. In preparing for productive thinking, students can also learn how to conduct research or inquiry, individually or in groups. This part of the foundation includes tools for stating a question or problem for research, collecting data, organizing and analyzing data, and presenting the results of an analysis.

Expression and Productivity Tools. Students can also learn to use a variety of tools and skills to express their ideas, such as making oral presentations, conducting a discussion, or debating. They can also learn to use productivity tools, such as computers, audio-visual equipment, scientific instruments, or other technical equipment specific to certain subject areas.

2. **Realistic Tasks.** This dimension of Figure 4.1 builds on the foundation skills and involves helping students to extend, apply, or use what they have learned about any topic in new situations. We refer to this dimension as "realistic" because it often involves applied tasks or challenges that are designed and presented by the teacher, and controlled or managed to fit the time and resources

available. Realistic tasks are exercises or activities that are interesting and plausible, but are not as open-ended or unpredictable as real life challenges. For example, watching a movie in which someone is flying an airplane might be a rather artificial experience when compared to really going up and flying a real plane. Practicing in a flight simulator (or using a good multimedia or "virtual reality" computer flight program) would be realistic. It is more engaging and plausible than just watching the movie, but also more easily managed and controlled, and quite a bit safer, than the "real thing." Solving story problems about money in a workbook might be quite artificial. Using play money to operate a "pretend" store in the classroom would be more realistic, and involving the students in actually setting up and operating a store at school might represent a real life experience. Experiences with realistic tasks help students put their own knowledge or competence to the test in an authentic way, build confidence, and heighten their motivation and involvement in learning.

Applications and Extensions of Learning. These include using case studies, games, role-playing, or simulation materials to give students opportunities to apply knowledge, information, or procedural skills they have learned. This level can also include laboratory exercises, competitive or cooperative team games, or other similar activities in which students "try out" something they have learned previously in a new or unfamiliar situation.

Practicing Problem Solving. When students have learned specific tools or methods for creative or critical thinking, problem solving, and decision making, it is helpful to give them some sample situations or practice problems for applying those skills. These experiences enable the students to use a deliberate, structured problem solving method to deal with a hypothetical problem or situation posed by the teacher. These often involve everyday life problems, but they can also be related to content. For example, in reading a story, you might stop when the main character encounters a problem. Have the students work in groups, using the problem solving tools they have learned, to define the problem and develop solutions as if they were the character. Then, they can read on and compare their solutions to the action taken by the character in the story.

Making Choices and Decisions. This set of realistic tasks involves giving students hypothetical situations or possible courses of action and asking them to evaluate the possible actions or responses they might take and the results or consequences of those choices.

3. **Real World Opportunities and Challenges.** This dimension of Figure 4.1 involves learning that takes place in real life, in which students actually carry out their plans or decisions and assume responsibility for the results and consequences of those actions.

Problem Solving. Real life problem solving occurs when students identify and work on a problem that they are actually facing—in the classroom, in school, on the playground, at home, or in the community—and carry their efforts through to a solution and a plan of action that they will actually implement.

Inventing. This involves asking students to identify a need or an opportunity for a new product or service and then to devise and produce a prototype or model of an invention that responds to the need. The students often participate in a fair or Invention Convention, in which their ideas and inventions are inspected and judged and in which they can present or explain the task they worked on and the product they developed.

Independent Projects. These can include individual or small-group projects in which the students select an area or topic, identify their own problem or challenge to work on within that topic, gather data, organize and analyze their data, and prepare a report or presentation on their results. It can include work done (by the students) for science fairs or other contests and competitions, independent study projects, or interest-based activity groups or clubs.

Practice Activity

Using the material you developed at the end of chapter 3, think about the productive thinking skills and tools from Figure 4.1 that you will incorporate into your unit of instruction. Examine the three dimensions in the circle in Figure 4.1 and the tools and skills described in this chapter. Which skills and tools are already familiar to you? How might you incorporate them into your content plan for the unit? Do you need to learn more about some of the other productive thinking tools and skills so that you will be able to use them in your unit, too? Which ones?

CHAPTER 5: PLANNING MATRIX

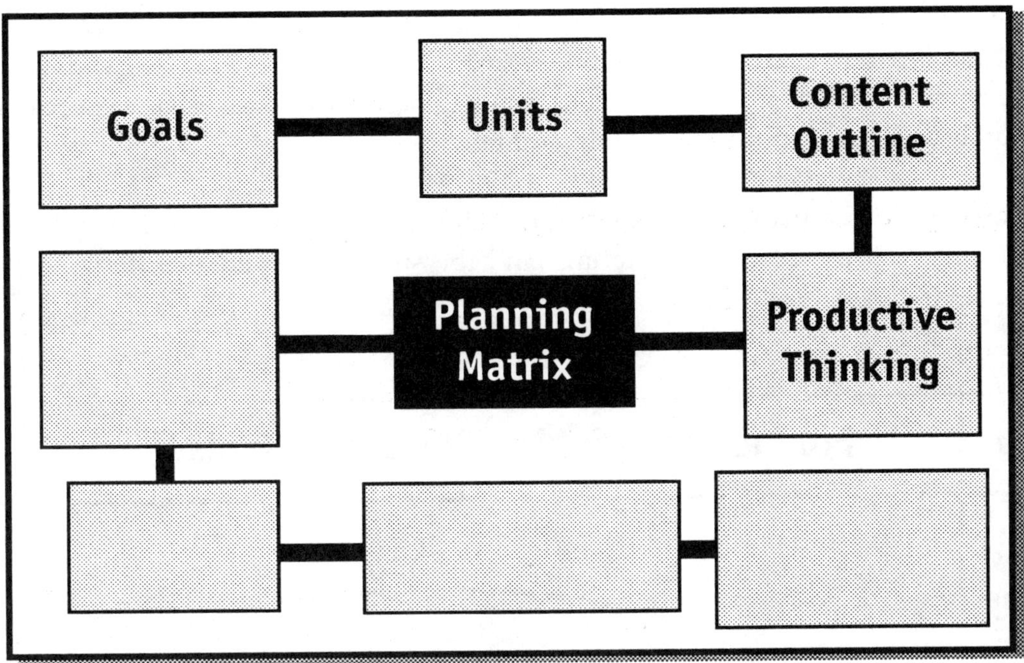

You should now have a content or concept outline with the topics you wish to include in your unit plan. You have reviewed the importance of productive thinking for instructional planning, and you are familiar with several productive thinking tools and skills that will be included in your unit. Your new challenge for this chapter will be, "How might I put these together to design and organize my instructional plan effectively?"

One good way to deal with this challenge is to use a matrix, placing the major headings of your content outline along one dimension and listing the important productive thinking dimensions along the other dimension, as shown in Figure 5.1. This will help you analyze each part of your content outline with each of the productive thinking dimensions you plan to include. An easy way to begin making the matrix is to use a sheet of newsprint or other oversize plain paper. Then, follow these steps:

1. Enter each of the major headings of your content outline at the left side (i.e., in the first column) from top to bottom. Leave a few inches at the top for process levels (see Figure 5.2).

2. Next, enter the list of thinking processes horizontally across the top of the page, beginning in the second column of the sheet. (Use the productive thinking skills and tools you have selected.)

3. Complete the horizontal and vertical lines to mark off each box or cell of the matrix.

4. Next, examine carefully each cell in the matrix. Begin with the first content topic and then check each thinking process for that topic. Ask yourself, "Can this productive thinking skill or tool be applied in this content area to describe some important learning for my students?"

5. Of course, not every topic will necessarily be combined with every thinking tool. However, by checking all possibilities, you will be certain to make these decisions deliberately and thoughtfully, not just accidentally.

6. Now you have a matrix that gives you an outline for planning your unit. It takes into account both the content of the unit and the student's productive thinking skills and tools.

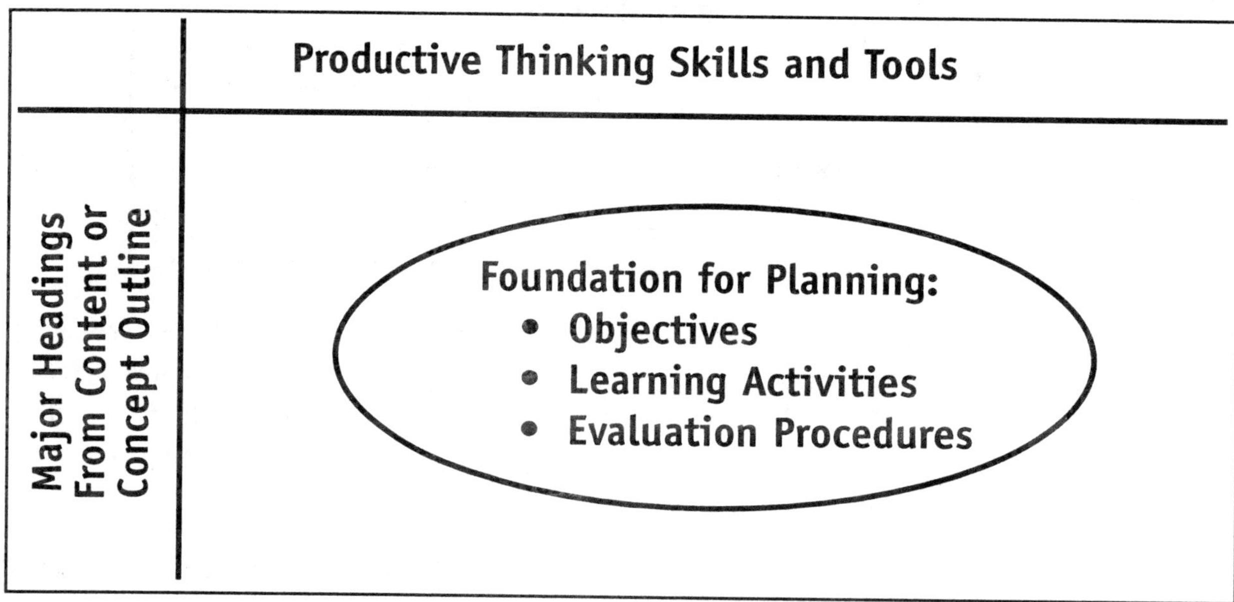

Figure 5.1: General Format for the Planning Matrix

Content Topics	Productive Thinking Dimensions		
	Foundations	Realistic Tasks	Real Life Opportunities

Figure 5.2: Sample Planning Matrix

You have also begun considering some other important questions about using your unit. Not every student will necessarily complete all the learning in any of these topics. The matrix is your "Master Plan" for the whole unit. It will guide you later in making decisions for guiding individual student learning. Several important concerns regarding individualizing and delivering your unit will be discussed in chapter 8.

As you continue to work through the unit planning process, you can use the matrix you've just completed for several different purposes. In the following chapters, you will learn to use the matrix to help you organize the objectives for each topic (chapter 6), the learning activities for each objective (chapter 7), and your record keeping and student evaluation procedures (chapter 9).

Suggested Activity

Create a planning matrix for a content or concept outline, using the productive thinking skills and tools you have selected. Then, identify one or more additional productive thinking tools that you had not previously considered and make some notes about how you might learn more about those tools and incorporate them into the unit in the future. What new questions, ideas, or themes might be considered if you were to extend or expand this unit in the future?

CHAPTER 6: OBJECTIVES

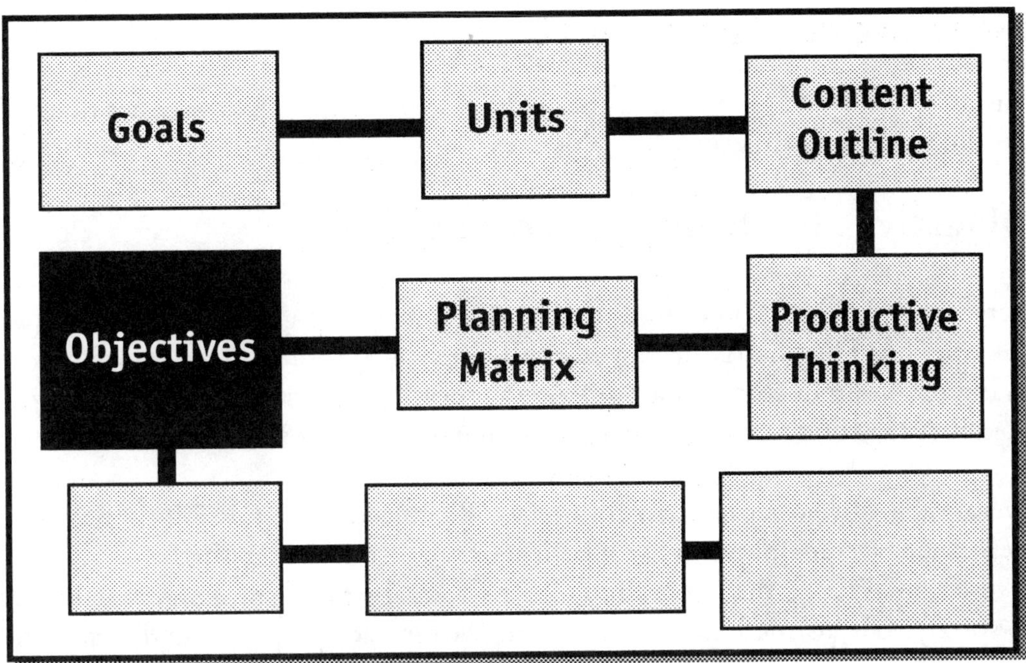

There's something very strange about the subject of instructional objectives. Almost every teacher who has been trained in the last two decades has taken at least one course—or probably several—in which how to write objectives was discussed. Many books have been written about how to write objectives. A great deal of research has been done on the effectiveness of explicitly stated objectives. Pressures for accountability stress the need to define precisely what is learned in school. But, despite all this concern, few teachers actually spend much time writing instructional objectives, and many teachers believe that written objectives are not necessary or useful in teaching.

There may be many reasons for this paradox. Written planning requires time that is not available to many teachers. Some courses fail to help teachers learn how to write objectives for more complex, important kinds of learning. Writing objectives is often carried out in isolation, rather than as an integral part of the process of instructional planning. Some approaches to writing objectives emphasize such great detail that the products may seem trivial or excessively wordy.

Characteristics of Useful Objectives

We can summarize several characteristics of useful objectives in this way:

1. A useful objective gives a clear statement of the important learning that should occur as a result of instruction.

2. The learning outcome describes student behavior as explicitly as possible.

3. A useful objective avoids language that is open to many interpretations (such as to understand, to appreciate, or to "really" appreciate).

4. A useful objective emphasizes what is important for the student to learn, rather than the teacher's behavior.

5. A useful objective states the kinds of observable performance or product that will show that the student has mastered the objective.

Using Objectives in Planning Your Unit

In chapter 5, we described procedures for creating a matrix to guide your efforts as you plan a unit of instruction. This matrix helps you analyze the way your content topics can be combined with important productive thinking skills and tools. But, at this point, the matrix has cells (specific combinations of a topic and a certain thinking skill) that look quite empty.

For several cells (but not necessarily for every cell), consider the possibility that the combination it represents is something that will be important—something worth the student's time and effort to learn. What is that "something"? In a formal way of speaking, that key question asks you to specify your instructional objectives, the learning activities you might use to help students attain those objectives, and the evaluation or documentation procedures that will enable you to know that the instruction was successful and that the objective was reached. What you will do, then, is "lift out" each cell in the matrix, one at a time. Decide whether it can be combined with an adjacent cell to create a worthwhile learning experience. While you don't want a few objectives that will be so broad that they are unclear and confusing, you also want to avoid having so many small objectives that they all begin to seem trivial or isolated and fragmented. When you determine that you have a meaningful cell, instead of a two-dimensional square, it will emerge as a cube, as shown in Figure 6.1.

For each cell in the matrix, then, a similar cube might be created (at least, the possibility exists for a cube for every cell). Working within the cube provides the cornerstone of the planning process. To be useful in teaching, objectives should be broad enough to encompass a worthwhile and meaningful learning experience and activities. The objective states the topic to be studied, the productive thinking process or skill to be used, and the product or outcome to be evaluated.

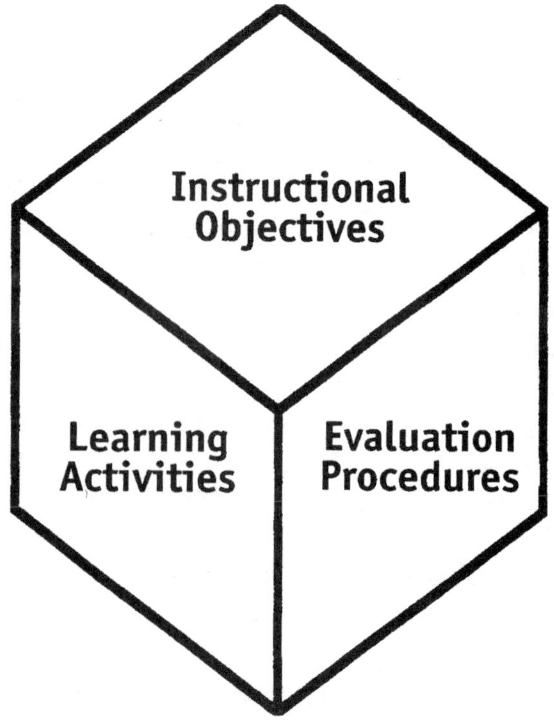

Figure 6.1: The "Planning Cube"

- **Instructional Objective.** This tells the important learning outcome for the particular combination of content and productive thinking skill or tool.
- **Learning Activities.** They describe what the students will do to reach the objectives.
- **Evaluation Procedures.** They specify the evidence you obtain to assess and document the student's performance or to verify that the student has attained the objective successfully. Here are some examples:
  - "In their study of colonial government, students will analyze the political structure of at least one colony, demonstrating the lines of authority in that colony."
  - "In their study of enzymes, students will locate information about the value of enzymes in human nutrition and summarize their findings."
  - "In their study of *Gulliver's Travels*, students will create a short modern version of the story in which they incorporate at least three contemporary political trends or issues."
  - "In their study of money and the economic system, students will generate a list of at least five different economic problems that might relate to shopping malls, and they will identify possible solutions for at least one of those problems."

The three parts of an objective are closely interrelated, of course. Begin with the content and the productive thinking skill or tool that is specified for the cell with which you are working. The product or outcome identifies the expected performance or result clearly. Including these parts will help you select or construct learning activities that will help students reach the objectives. We will work on learning activities in the next chapter. The evaluation procedure documents the learner's successful work (learning or achievement of the objectives). Evaluation will be the topic of chapter 9. As you proceed through this book, subsequent chapters will help you design learning activities and evaluate the outcomes. But, in this chapter, our task is to consider just the first, basic step: completing the matrix with objectives that take into account both content and productive thinking.

Writing Objectives for the Productive Thinking Dimensions

It is easy to assess whether an objective is relevant for the content topic that has been selected. If you are working on the topic, "Identifying the Colonies," for example, the objective must have something to do with that subject. It probably deals with listing, naming, locating, or describing the colonies. An objective dealing with life and experiences of young people in colonial schools or with the political process of governing the colonies wouldn't fit in this topic, even though it might be relevant and useful in another topic elsewhere in the unit. It is not always as easy to ensure that your objective fits with the productive thinking dimension in the matrix. Keep in mind the kinds of learning that are involved in each of the productive thinking dimensions. If you aren't clear in your own mind about what kind of student behavior might represent each of the productive thinking dimensions, review chapter 4. The main emphasis of each of the three basic dimensions of our framework for productive thinking instruction is summarized in Figure 6.2.

This dimension of our framework for productive thinking instruction:	Emphasizes:
Foundations	
Knowledge	The information students need.
Generating Tools	Producing many, varied, or unusual ideas.
Focusing Tools	Analyzing, refining, organizing, developing, evaluating, judging, or choosing ideas.
Research and Inquiry Tools	Defining a problem, collecting data, organizing and analyzing the data, and presenting results.
Expression and Productivity Tools	Using equipment and materials to support your thinking and being able to present results or ideas.
Realistic Tasks	
Application and Extension	Using what you have learned in new or different situations or tasks.
Practicing Problem Solving	Applying problem solving methods to sample or hypothetical problems.
Making Choices and Decisions	Using criteria to choose or decide and justifying your conclusions.
Real Opportunities and Challenges	
Problem Solving	Working on problems and carrying out the solution or action plan.
Independent Projects	Defining a topic or area of interest, pursuing it in-depth, and completing a report, project, or presentation.

Figure 6.2: Dimensions of the Productive Thinking Framework

Chapter 7: Learning Activities

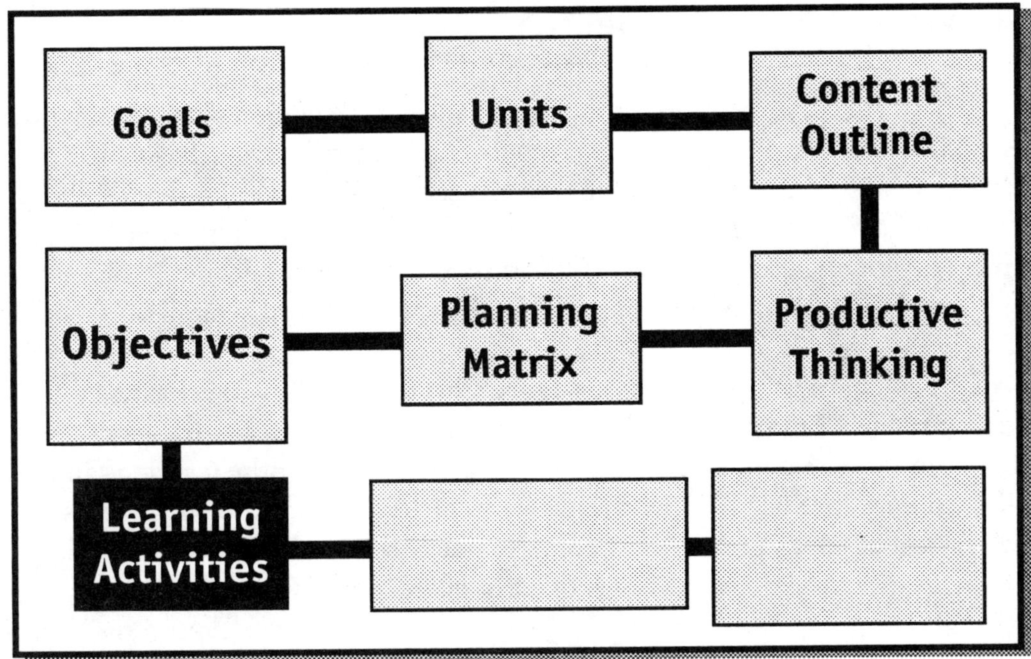

A well-planned instructional unit provides many different alternatives for student learning. As we emphasized in the previous chapters, two important aspects of instructional planning are for students to have opportunities to use and develop different thinking skills and for them to participate in instruction for which there are clearly defined objectives.

But, "objectives for different thinking processes" tells only part of the story. Providing many alternatives for learning also requires that students have opportunities to engage in a variety of different learning activities. A learning activity is anything done by a student—or a group of students—to enable them to meet an objective. It is what the students actually do. As you can see, then, variety in thinking processes and variety of learning activities are closely related.

You probably noticed, when you were writing objectives for your matrix, that objectives for different thinking processes involved different kinds of activity or behavior by the students. The objectives lead you naturally to the next step in the planning process: developing learning activities for each of the objectives in your matrix. There are two important things to keep in mind as we begin this step:

1. Your students bring with them to the classroom experience, maturity, skills and ability, and different learning styles. Therefore, your unit can be effectively individualized if you can plan several different activities for reaching a specific objective.

2. Objectives planned for different thinking skills will also involve a variety of different activities. Some activities are particularly well-suited for developing certain thinking processes.

Many Activities for Each Objective

Educators have talked for years about the importance of "recognizing individual differences." Individualized instructional planning helps teachers take such talk seriously and accomplish that goal realistically. Let's look at an example of how you can take individual differences into account by planning many activities for the objectives in your matrix. Think about a knowledge objective for the topic, "Identifying the Colonies." One simple example would be: "After studying the colonies, the student will be able to name and identify the 13 original colonies."

What might you have your students do to meet this objective? If they haven't yet learned to name and identify the 13 original colonies, what kinds of tasks or activities would enable them to do so? For example, you might pass out copies of a map, showing the locations and giving the names of the colonies for your students to use in memorizing that information. As you design a unit, it will be a good idea to try to list more than one possible activity for each objective, since there are many possible ways that any objective can be taught and learned.

In addition to the map activity, you could proceed in several other ways. Keep in mind that, since this is a *knowledge* objective, we are concerned with the student's knowing and remembering the names and locations of the 13 original colonies. How else could we accomplish this? In the following spaces, write down at least three more learning activities to help your students (individually or in groups) remember the names and locations of the 13 original colonies.

1.

2.

3.

Remember that, for each objective, you might have several activities that are alternatives for students to carry out in mastering or achieving the objective. Here is an illustration from the "Realistic Tasks" dimension of productive thinking and the content area of family needs. It involves comparing and contrasting ideas.

Objective: The student will be able to contrast and compare different types of family needs.

Activities: Compare the financial situation of an inner city family and a wealthy suburban family; or compare the food and clothing needs of American and Mexican children.

Here are 10 other possible learning activities for our knowledge objective on naming and identifying the 13 original colonies. Compare these to your responses on the previous page:

1. View a film or filmstrip.
2. Listen to a taped presentation.
3. Work a puzzle with pieces representing the 13 colonies.
4. Develop a complete, labeled map from an outline map.
5. Complete a crossword puzzle on colony names.
6. Work on a bulletin board matching names and locations.
7. Use flash cards with students working in pairs.

8/9. Use the "team learning" or "circle of knowledge" strategies described by Dunn and Dunn (1978).

10. Use a game to teach and practice the names of the colonies.

Keep in mind that this objective only requires that we help students remember the names and locations of the 13 colonies. Activities for other, more complex objectives come later. This exercise should help you recognize that many different kinds of learning activities can be planned for any of your objectives. It should also remind you of the importance of providing alternatives for students to use in achieving any objective.

Important Note—Distinguishing Objectives From Activities

Keeping straight in your mind that the important difference between Objectives and Learning Activities can be complex—but, it is important.

Objectives:	**Learning Activities:**
State the **goal** or major purpose that you hope to accomplish.	Describe the **means or methods** for reaching the objective.
Tell **what** you hope the students will know or be able to do as a result of the instruction.	Tell **how** the students will learn that.
Example: Know the location of the original 13 colonies.	**Example:** Label the 13 colonies on a blank map.

Different Activities for Each Productive Thinking Dimension

The second important thing we stressed at the beginning of this chapter was that different learning activities can be used to assist you in fostering productive thinking by your students. This is easy to see when we consider some extreme examples. If you are concerned with developing *creative thinking skills*, for example, reviewing flash cards or copying and memorizing a list of definitions would not be appropriate learning activities. By the same token, a very open-ended "brainstorming" session would not be a very effective way to teach students how to spell a vocabulary word correctly. In this chapter, we will explore some of the kinds of learning activities that would be appropriate to consider for each of the three dimensions of productive thinking (foundations, realistic tasks, and real life opportunities and challenges).

Learning Activities for the Foundations Dimension

The foundations dimension represents the knowledge and information that students need and the thinking and working tools that will also help them work successfully on realistic or real life tasks and challenges. Figure 7.1 presents a list of several examples of appropriate learning activities that can be used effectively in this dimension.

Learning Activities for Information and Knowledge Base

Many well-known and widely used learning activities are effective in helping pupils acquire, organize, and understand information and build their knowledge base in any content area. Some of these learning activities include well-planned lectures or presentations, question-and-answer sessions, recitations, and workbook activities requiring children to read and then answer questions regarding the content. Since this part of the foundation stresses the students' ability to remember and understand information or procedures, activities that provide opportunities for review and practice are often useful. Games and puzzles in which students earn "points" or advance to a goal by answering simple questions or performing tasks correctly can be used to provide practice and maintain student interest and motivation. Answering thought questions fosters understanding. Programmed instructional materials, interactive multimedia programs, or computer-assisted instructional programs may also be useful, and many good programs are available. Some of these may be too slow-paced for bright, active learners, however, and may become "dull" or uninteresting if they do not allow for flexible use or choices for students. Computers can help make it possible for students to make rapid progress or skip over material they already know well. Information searches, which may take many forms (including the common "scavenger hunt" approach, where each student has a list of facts or information to seek), can also be used effectively.

Keep in mind these principles when working with knowledge and information activities:

1. Organize the material to be learned into meaningful units or segments.

Illustrative Learning Activities for the Foundations Dimension				
Knowledge and Information	**Generating Tools**	**Focusing Tools**	**Research and Inquiry Tools**	**Expression and Productivity Tools**
Participating in question and answer sessions	Brainstorming; open-ended questions or tasks; brainstorming variations	Observing carefully	Learning and practicing library research skills and search tools	Doing data organization and presentation activities
Completing workbooks or activity sheets	Responding to "What if …" or "Just suppose …" questions	Making inferences	Posing or constructing research questions or hypotheses	Using listening skills
Using computer-assisted instruction, computer games to practice skills	Identifying gaps or missing elements	Reaching logical conclusions and making deductions	Narrowing or broadening research questions; "Ladder of Abstraction"	Creating and using charts or graphs
Working on board games, team games, or puzzles	Identifying paradoxes or puzzling questions	Selecting relevant data or attributes	Interviewing	Using oral presentation skills (planning, delivering)
Doing word or information searches; memory games or quizzes	Improving products	Classification of objects or making clusters of ideas ("Hits/Hot Spots")	Making and using surveys, questionnaires	Using audio-visual tools or equipment
Making and using mnemonic devices	Doing story or picture completion; making up unusual titles	Outlining and identifying main ideas	Tallying, charting data	Using computers or other technologies to organize and present ideas
Looking up definitions	Using SCAMPER or other idea checklists	Sequencing	Doing descriptive statistics (central tendency, variation, correlation)	Using team work and collaboration skills
Using flash cards	Force-fitting; using many senses to make new connections	Solving logic or mystery puzzles	Doing inferential statistics and hypothesis testing	Using leadership behaviors or practices
Participating in "circle of knowledge" activities	Listing attributes	Reasoning by analogy	Applying qualitative analysis methods	
Doing team-learning activities, peer teaching, or tutoring	Combining possibilities with the morphological	Recognizing errors of logic or fallacies in reasoning; identifying persuasion or propaganda	Interpreting data; generalizing; recommending	
Stating ideas in own words; giving examples; paraphrasing	Identifying changes	Summarizing information		
	Identifying and using metaphors	Generating and selecting criteria		
		Identifying ALoU		

Figure 7.1: Illustrative Learning Activities for the Foundations Dimension

2. Provide for frequent review and "breaks" to change the pace after difficult lessons.

3. Use clear illustrations or examples that are within the students' experience to improve recall and understanding.

4. Use mnemonic devices or "memory aids" to help students remember groups of facts. You can also encourage the students to create their own memory aids.

Also, keep in mind that a powerful knowledge base involves more than just remembering facts and information. It stresses the ability to understand what is learned, to make it part of one's experience. Learning activities for Knowledge and Information should also provide the students with opportunities to explain or relate, to share, to give their own examples or illustrations, or to make predictions and estimates.

Learning Activities for Generating Tools

Many learning activities can also be used to help students learn and use tools for generating ideas. These tools are an important part of the foundation, since they help lay the groundwork for creative thinking and problem solving. To help your students learn and use the generating tools effectively, you will also need to help them learn to defer judgment, to try to give many responses (when you pose questions that are open-ended), and to ask questions, not just to assume that their only task is to give answers to the questions you ask. You should seek ways to use several of the generating tools as part of your day-to-day instruction. Creative thinking should not be merely "play time" nor "fun and games," although you and your students will enjoy doing it. You should also remember that creative thinking can be an important part of your students' learning, in almost any subject area. Many of the specific learning activities in Figure 7.1 may be things you haven't heard of before. Some are likely to involve tools you have used, even if the names are not familiar. Rather than discuss each one in detail you can use these sources for additional information:

Treffinger, D. J., Isaksen, S. G., & Dorval, K. B. (2000). *Creative problem solving: An introduction* (3rd ed.). Waco, TX: Prufrock Press.

Treffinger, D. J., & Nassab, C. A. (2000). *Thinking tools lessons*. Waco, TX: Prufrock Press.

The December 1996 issue of *Think™ Magazine* is a special issue devoted entirely to explaining and giving classroom examples of both generating tools and focusing tools.

Learning Activities for Focusing Tools

The focusing tools are also important in preparing students to be critical thinkers, to deal effectively with a variety of real life situations, and to be successful in working on more complex academic activities and projects. Learning activities for the focusing tools involve such skills as: being observant, categorizing or classifying ideas and information, thinking inductively and deductively, to making basic comparisons and using analogies in reasoning, determining whether data are relevant or irrelevant, and making appropriate inferences from information provided. A number of useful resources can help you plan appropriate activities for teaching focusing tools as part of your unit. The resources listed above, for the generating tools, also include relevant material on focusing tools. Several other useful resources for focusing tools are:

Black, S., & Black, H. (1984). *Building thinking skills.* Pacific Grove, CA: Critical Thinking Press and Software.

Draze, D. (1981). *Connections.* San Luis Obispo, CA: Dandy Lion.

Harnadeck, A. (1978). *Mind benders.* Pacific Grove, CA: Critical Thinking Press and Software.

Learning Activities for Research and Inquiry Tools

The research and inquiry tools provide the basic skills that serve as the foundation for planning and conducting original inquiry or research. At the foundations level, your emphasis will be on helping students learn what the research and inquiry tools are, how to use them, and when they might be appropriate to use. You do not expect the students at this stage to be able to plan or carry out original research or inquiry projects, but simply to learn the tools they will need to do more complex practice activities or original projects later. Appropriate exercises at the foundations level will usually focus on a specific tool and provide a basic activity for the students to do to enable them to understand the tool and its purposes. This column in Figure 7.1 includes a number of tools that vary in complexity or difficulty. For the activities you plan, it will be important to select the tools that will be appropriate for the age, ability, interests, and previous experience of the student(s) with whom you are working.

Learning Activities for Expression and Productivity Tools

The expression and productivity tools provide opportunities for students to learn how to express ideas, processes, or products in tangible forms—products or expressions that might involve written words, oral presentations, numerical or mathematical expressions, visual representations, or other ways of expressing themselves. The productivity tools include the appropriate and effective use of technology or equipment and the personal or interpersonal skills needed to enable people to use and share their knowledge and ideas in many ways. Many definitions of creativity or productive thinking emphasize the importance of being able to communicate ideas to others and present your thoughts effectively in ways or formats that are appropriate to your interests, personal strengths, talents, and learning style preferences.

Learning Activities for the Realistic Tasks Dimension

The realistic tasks dimension represents opportunities that enable students to apply or extend the knowledge, information, and tools they have learned in the foundations dimension, to practice problem solving methods, and to learn how to make effective choices and decisions.

The activities for this dimension should be engaging (motivational or "high interest") to the students, but they will usually be defined and presented (or at least coordinated and supervised) by the teacher. That is, realistic tasks can be managed or controlled, in relation to time, resources, working activities, and outcomes. Unlike real problems and challenges, for which the students' work may take them wherever the problem requires them to go, when you are working on realistic tasks, you can maintain a high level of structure and direction for the students' work. In realistic tasks, the teacher also selects or constructs tasks in specific content areas or on certain topics so the students' work can focus on issues, themes, and content that is relevant to one or more curricular areas that will be important and appropriate for the students to work on. Some realistic tasks involve "everyday life" experiences that are within the students' general knowledge base and maturity. These tasks may not necessarily be linked specifically to a particular content area or subject matter topic.

The realistic tasks level involves efforts to engage students in what has been described as "authentic student performance." That is, the major purpose of the learning activities at this level is to enable the students to use or extend the knowledge, information, and tools they have learned, providing tasks that are similar or comparable to those encountered by people in the actual course of life—at home, in the community, at work, or at leisure. It may be helpful to remember three "C's" that are important when students are working on realistic tasks:

- **Competence**. These tasks help students understand and use what they know and the tools they have learned.

- **Confidence**. Realistic tasks engage students in learning activities that can be planned, managed, and directed carefully to foster success and build students' confidence in their productive thinking.

- **Commitment**. Realistic tasks serve as a starting point for students to explore topics, themes, issues, or problems that they recognize as important and worth pursuing.

Figure 7.2 presents a list of several examples of appropriate learning activities that can be used effectively in engaging students in realistic tasks.

Illustrative Learning Activities for the Realistic Tasks Dimension			
Applications or Extensions of Learning		Practicing Problem Solving	Making Choices and Decisions
Doing simulation and extended gaming activities	Participating in peer tutoring, peer teaching, or cooperative learning groups working on exercises or project tasks assigned by the teacher	Doing role-play, role reversal, or socio-drama activities	Generating criteria for questions, decisions, or tasks presented by the teacher
Producing newspaper ads, stories, or brochures	Doing "Morning Talk" or "Show-n-Tell" presentations to the class	Working with practice problems provided by the teacher	Using criteria to make judgments, choices, or decisions about a case or situation presented by the teacher; justifying the choice, judgement, or decision
Producing simulated radio or TV ads	Making predictions or estimating from a given set of data	Participating in group problem solving programs (Future Problem Solving, Destination ImagiNation)	Given specific criteria, evaluating several possible conclusions or decisions, and justifying the choices
Building models	Constructing charts, graphs, or diagrams to represent given data or relationships	Making up problems with given sets of data (e.g., using many math facts in constructing story problems)	Identifying possible causes, consequences, and implications of events or decisions
Preparing dioramas or displays	Debates on assigned topics	Identifying and solving problems encountered by characters in a story	
Creating multimedia computer programs	Holding panel discussions on specific topics	Using historical situations or events as data for problem solving	
Making oral, visual, musical, or dramatic presentations on a topic	Comparing and contrasting	Identifying problems, gaps, or paradoxes in a new situation presented by the teacher	
Conducting procedures (e.g., laboratory exercises) or simple experiments provided by the teacher	Using webbing, idea trees, or concept mapping for a given topic		
Making up classifications or classification systems			
Applying information or tools for case studies or exercises provided by the teacher			

Table 7.2: Illustrative Learning Activities for the Realistic Tasks Dimension

Learning Activities for Applications or Extensions of Learning

Activities for applications or extensions of learning should give students opportunities to use their knowledge, information, and ideas in new situations—in working on structured projects or practice problems they have not encountered before. The problems and challenges must be new for the students, of course, or else the level of the activity is merely additional drill and practice at the knowledge level. The application or extension activities take place when the teacher identifies a project or problems that will be of interest to some or all of the students, building upon the basic information and knowledge base that the students have learned.

Let's consider an example of an activity in this area, drawing once again on the Colonial America unit: "The students should be able to use information about the colonies and their location in a new context or setting." One learning activity might be: "Write a travel brochure advertising the location and attractions of a colony of your choice or for a tour through the colonies." Another might be, "Suppose you wanted to attract more visitors to your colony. Make a poster containing a collage of drawings or pictures that portrays the attractiveness of your colony and its location." These represent applications or extensions of learning, since the student is using his or her knowledge and understanding of the colonies to work on the new activity. In this dimension, keep in mind that there should be more for the learners to do than merely recall or repeat something they have learned. There should be an opportunity for them to put their learning to use actively in a new situation.

Realistic tasks also involve activities that ask students to analyze ideas and information or to compare, contrast, or identify the important parts or characteristics of a problem or situation. You might begin by asking students to examine the parts of a single object or situation. For example, this can involve using a foundations tool such as attribute listing. Consider the ordinary telephone. If we wanted to improve this instrument, we might begin using attribute listing by asking, "What are the major parts of a telephone?" The students might list such attributes as: "communication device with earphone or speaker, microphone or transmitter, dial, bell that rings, plastic body," and so forth. Then, each of those attributes might be used as the starting point for improving the phone and might lead to more complex activities, as well (comparing and contrasting phones, studying how telephone communication has changed and is still changing today, or others).

Learning Activities for Practicing Problem Solving

There are many ways to help students practice using structured methods or processes for problem solving. Any story the students are reading can be a starting point for problem solving. What problem is the character facing? What should he or she do about it? Engage the students in defining and solving the problem from the character's point of view before they read how the author solves the problem. Historical events or situations can also be a good starting point for practicing problem solving in groups. There are also a number of published resources that provide practical, manageable problems for students at different ages. These include:

Draze, D. (1986). *Primarily problem solving*. San Luis Obispo, CA: Dandy Lion.
Draze, D. (1994). *Creative problem solving for kids*. San Luis Obispo, CA: Dandy Lion.
Draze, D. (1994). *Pickles, problems, and dilemmas*. San Luis Obispo, CA: Dandy Lion.
Eberle, B., & Stanish, B. (1996). *CPS for kids*. Waco, TX: Prufrock Press.
Elwell, P. (1993). *CPS for teens*. Waco, TX: Prufrock Press.
Isaksen, S. G., Dorval, K. B., & Treffinger, D. J. (2000). *Creative approaches to problem solving* (2nd ed.). Dubuque, IA: Kendall/Hunt.
Stanish, B., & Eberle, B. (1997). *Be a problem solver.* Waco, TX: Prufrock Press.
Treffinger, D. J. (2000). *Creative problem solver's guidebook.* (2nd ed.). Waco, TX: Prufrock Press.
Treffinger, D. J. (2000). *Practice problems for creative problem solving* (3rd ed.). Waco, TX: Prufrock Press.
Treffinger, D. J., Isaksen, S. G., & Dorval, K. B. (2000). C*reative problem solving: An introduction* (3rd ed.). Waco, TX: Prufrock Press.

Programs such as FPSP (Future Problem Solving Program; www.fpsp.org) or Destination ImagiNation (www.dini.org) also provide excellent opportunities for students to practice and apply problem solving methods and tools.

Learning Activities for Making Choices and Decisions

This dimension involves helping students practice the important skills of judging or evaluating, making choices, and making decisions. Activities in this group include exercises in which students learn about the nature and use of specific criteria and practice applying criteria to cases, tasks, or scenarios provided by the teacher. Evaluation, in the sense in which we are using it here, involves the students' ability to make judgments and decisions. We are not talking about how you will evaluate the students' performance or learning, which will be considered in chapter 9. Making choices and decisions gives the students opportunities to learn and practice the role and the skills of the judge or the decision maker. For any content area, there can be many opportunities for students to learn how to make critical judgments or decisions. Activities for evaluative thinking involve the student in such responsibilities as:

- working in groups to determine (and, subsequently, to apply) criteria for judging projects or presentations;
- evaluating the strengths and weaknesses of their own projects or solutions;
- making decisions about the consequences of hypothetical events or situations;
- using established criteria to decide among alternatives or competing choices and explanations; and
- assessing the correctness of statements or conclusions.

We often speak of "showing good judgment," "making responsible choices," or "decision making" as goals of education. These activities help students use evaluation and decision making processes effectively in their day-to-day instructional content and, thereby, help in accomplishing such goals.

Learning Activities for the Real Life Opportunities and Challenges Dimension

The real life opportunities and challenges dimension represents the most complex level of our framework for productive thinking. At this level, students are not just practicing or doing exercises. They are engaged in activities that extend beyond getting a grade or praise from the teacher or completing a project or paper that is displayed on a bulletin board (or a refrigerator door). Real life opportunities and challenges involve activities and experiences that are important, or even "passionate" for the students, and for which the consequences matter in their life. Guiding learning activities at this level demands careful planning and a high degree of imagination and original thinking. These activities also involve helping the students learn to be

Illustrative Learning Activities for the Real Life Opportunities Dimension

Problem Solving, Inventing, Independent Projects

Producing an original plan for a project, experiment, or investigation related to academic or vocational areas and interests. Carrying out the plan.

Defining and solving a real life problem of personal concern selected by the student. Dealing with real problems in the classroom, school, or community. Taking action to implement the solution(s).

Organizing and carrying out an original research study.

Planning, organizing, and carrying out a special activity or event (student-defined).

Inventing an original formula, process, procedure, or product in any copyright or patent category. Participating in an Invention Convention. Producing and disseminating the product.

Entering an academic contest or competition that involves original student work

Participating in projects or programs that involve creating and operating a business (e.g., Junior Achievement).

Planning and carrying out service project(s).

Planning, creating, and presenting work in any visual or performing arts area.

Table 7.3: Illustrative Learning Activities for the Real Life Opportunities Dimension autonomous, independent, or self-directed learners. Teachers cannot be there with every student, throughout their lives, to tell them what to do in each new situation. Eventually, each student must be able to engage in complex reasoning and synthesis in order to become an effective or successful individual. Figure 7.3 presents examples of a number of learning activities that represent real life opportunities and challenges.

A Note About Sequence

We have presented the three dimensions of the productive thinking framework as if they should, or do, occur in a specific or fixed sequence, from the foundations, to the realistic, and then eventually to the real life opportunities. In fact, instruction does sometimes follow such a sequence, but it does not always or necessarily do so. There may be many valid teaching and learning experiences that start when people are actively involved in real problems and challenges, and then involve the need for new knowledge, information, or tools. Do not assume, then, that you must always plan and carry out learning activities at every level, in every category of any of the levels, or in a certain prescribed or fixed sequence.

Exercise: Developing Learning Activities

Now you should consider your own unit plan. So far, you have a set of objectives that you developed from your matrix. Your next step is to review each (or any, but not necessarily "all") of the "cubes" or "cells" in your matrix. Each cell might contain at least one objective (or perhaps several) that combines a specific content topic with one of the productive thinking dimensions. Your job now is to devise the learning activities that will be best-suited to help students attain the objective(s). In doing this, the following steps may be helpful.

1. Look at one "cell" of the matrix at a time. Thus, you'll begin in the upper left corner of the matrix, working your way across the row before moving down to the next row.

2. As you begin each row, study the topic for that row so the appropriate content is clearly in mind.

3. Then, for each cell, consider the productive thinking dimension and the specific skills within that dimension that the cell includes. Review some of the kinds of learning activities that are well-suited for that level.

4. Think of possible audio-visual or technology resources you can use on an individualized or small-group basis to help students learn. For example, they can view slides, videotapes, or films or work on computers at specific work areas or you can set up in your classroom.

5. Provide opportunities for your students to develop audio-visual materials as a part of, or as the main focus of, the presentation of completed projects. They can create slides,

overhead transparencies, videotapes, or multimedia computer programs to present the results of their projects. The media specialist in your school can help in guiding students in these projects.

6. Plan for group project activities for students who wish to work together. By pooling their knowledge and creative capacities, all the students in a group can carry out a project that might be much more effective than if each of the students had worked alone.

7. Try to develop several possible learning activities for each objective. Remember, the test of a good learning activity is: "If the learner does this, will it help him or her reach this objective?"

8. If your objectives were written in too narrow or specific a form, you may see only one learning activity that is appropriate. When this occurs, review Figures 7.1 through 7.3 (as appropriate) to consider other possible kinds of activities that might be useful. If you find some, they should guide you in revising your original objective to make its wording more inclusive or accurate.

9. In some cases, you may decide that certain activities must be done by all students to assure some basic learning. Such activities can be marked with an asterisk, and the requirement can be presented in writing to the students.

10. Remember that not every student in your class will do all of the activities, nor meet all of the objectives, in your plan. The reason we are trying to create many possibilities is to create a rich set of alternatives that can be selected for various students. Thus, it is important now to include as many possibilities as you can (and to remember you can always add more ideas later).

11. It's also important to remember here that "learning activities" are not only things you make up on your own. For any objective in your matrix, you should feel free to use any published or shared materials you can locate and obtain. Your learning activities should involve all the resources you have available to help students attain the objectives efficiently and happily.

Before you begin the next chapter, work on planning several learning activities for each of your objectives. Be certain to include several alternatives for each objective and be certain your activities really involve the appropriate productive thinking dimensions and skills. After you have written your objectives, get some feedback by reviewing and discussing your plans with a colleague.

Chapter 8: Individualization & Delivery

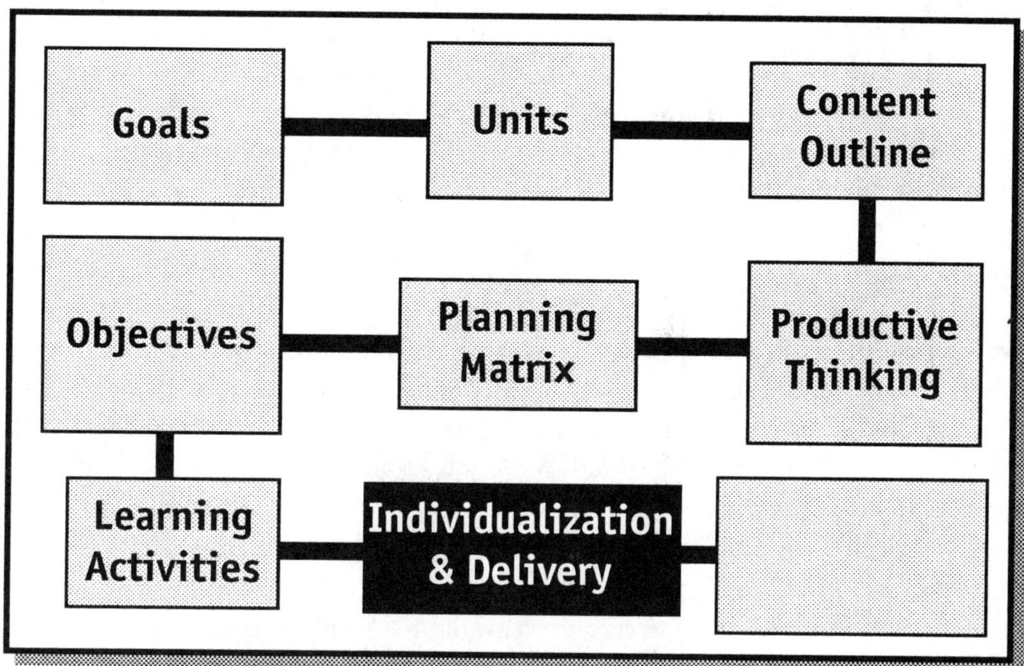

In good instructional planning and curriculum development, educators recognize that students do not all learn in the same way and that, at any given time, all the students in the class are not likely to "need" the same instruction.

Yet, "individualizing" is more honored in theory than in practice. While many acknowledge the importance of differences among people and their consequences for learning and instruction, putting those concerns into practice has been elusive, difficult, and often frustrating for teachers.

In some classrooms, efforts at "individualized" instruction merely change the order or timing of assignments, and all students eventually do all of the same assignments and activities. The only thing that is really individualized is the rate or time at which the assignments are undertaken. We believe that a more effective approach makes it possible for various students to be provided with different activities, different thinking processes, and different content or subject matter. Such an approach takes into account all the factors in Figure 4.1: the three dimensions of productive thinking, personal characteristics of the students, metacognition, and the context for learning. In working with the content of chapters 4 through 7, you have planned a content outline or web, constructed a matrix, developed objectives, and planned a variety of learning activities. How will you now prepare to deliver (or teach) this unit? How will you be certain that you are individualizing, and taking all the factors in the model in Figure 4.1 into account? These are the questions this chapter will address.

Personal Characteristics

One way to consider individualization in the process of instructional planning or curriculum design is through incorporating breadth, depth, or variety into the content or including a wide range of topics that will span differences in age, ability, background, and interests. There may well be a common core of material that is important and necessary for all students to learn. However, we try to respond to individual differences by including a wide range of topics, resources, outcomes, activities, or projects. There is also another useful way to approach the challenge of individualizing. It is represented in the question below.

"Who Learns What?"

When you take a unit you have planned following the guidelines in this book into a classroom and prepare to teach the unit to your students, one of the first problems you will probably encounter is, "Will every student be expected to meet all of those objectives or do all of the activities?" Of course, you will be able to tell, merely by looking at your plan, that it includes more possibilities and choices than you could ever reasonably expect all of your students to do. Some parts of your unit plan may be useful in working with your entire class at certain times and for certain purposes. But, you must also have a plan that will help you to consider the different needs of various students and the resources available to help you meet those needs effectively and efficiently.

Assessment programs that use test data for diagnostic or prescriptive planning, for example, can be one resource teachers use to determine their students' actual skills, needs, or instructional levels. Careful analysis of diagnostic assessment data and considering the specific areas of strength or weakness (and not merely global indicators such as percentiles or grade level equivalents) have often been used to help teachers plan specific instructional strategies for initial learning, supplementary practice, and remediation. They are also important and useful for enrichment (e.g., Renzulli, 1994) or for acceleration (Stanley, 1980) for capable students. Simple pre-testing on the content of a new unit of study, or an informal teacher-made skills checklist used in conjunction with observations of students' performance can also be helpful ways to deal with the issue of "who learns what."

Other teachers use checklists or rating scales to learn more about their students' characteristics and needs. While these data are much less precise and are, most likely, not specific to a certain content or subject matter area, even informal knowledge of the student's personal characteristics can help teachers understand the learner's strengths, limitations, and needs.

In order to select from among the many alternatives in a carefully designed, comprehensive unit plan, however, you may find it helpful to obtain some additional information about your students. For example, you will want to take into account their different learning styles or preferences. Fortunately, such decisions need not be made by guesswork. Several instruments are available that enable you to obtain a more complete picture of your students' learning styles and preferences.

For more information on these topics, see:

Dunn, R., & Griggs, S. (1988). *Learning styles: Quiet revolution in American secondary schools.* Reston, VA: National Association of Secondary School Principals.

Lawrence, G. (1984). *People types and tiger stripes* (Rev. ed.). Gainesville, FL: Center for Applications of Psychological Type.

Renzulli, J., & Smith, L. (1978). *Learning styles inventory.* Mansfield Center, CT: Creative Learning Press.

Renzulli, J. S., Smith, L. H., & Reis, S. M. (1982). Curriculum compacting: An essential strategy for working with gifted students. *Elementary School Journal, 82,* 185–194.

Metacognition

Metacognition is a very important factor in your efforts to teach any unit in ways that will help each learner be successful and autonomous. *Metacognition* is a technical term for the skills that are involved in monitoring, managing, and modifying your own learning and in being aware of your own thinking and learning processes and activities as they take place. Metacognition involves helping students to ask (and to respond to) such questions as:

- What am I going to do, how am I going to do it, and why should I do it that way?
- What am I doing now, how well is it working, and should I keep doing it, stop, or modify what I am doing?
- What did I do, was it successful, and how might I do it differently or better in the future?

Some of the more specific issues and aspects of metacognition are summarized in Figure 8.1.

The Context for Learning

Context refers to the psychological and physical setting in which teaching and learning take place in the classroom. The *physical setting* refers to the way you set up or organize and use the furniture, equipment, and materials in a classroom. These factors can have considerable impact on the kinds of learning activities that will be able to take place in the classroom, and they may be related to the opportunities that students can have to apply productive thinking, too. The materials on teaching and learning style by Dunn and Dunn (1978), mentioned previously in this chapter, include many practical suggestions for designing (or redesigning) a classroom to ensure that a variety of learning styles and preferences can be accommodated. Their suggestions will also help you to consider ways to arrange, or rearrange, your space and facilities to encourage students to use productive thinking skills, whether they are working alone, in peer or small groups, or as an entire class. Think of your classroom as a "laboratory" for exploration, investigation, discovery, and concentration or study—not as a place that is only for all the students to engage in a fixed set of activities throughout the period or the day.

The second major aspect of the context for learning involves the *climate*—the psychological setting or the way people feel about the physical and interpersonal setting of the classroom. The climate of the classroom can encourage productive thinking, or it can stifle and inhibit it. Research has shown that nine general factors make up the "climate for creativity and innovation" in a group (e.g., Isaksen, Treffinger, & Dorval, 1996). These are:

Challenge & Involvement. The degree to which students are involved in the daily operations, long-term goals, and in sharing a vision of the classroom as a learning environment.

Freedom. The extent to which students have the opportunity for independence in their behavior.

Trust and Openness. The extent to which the students feel there is emotional safety in their relationships with each other and with the teacher.

Idea Time. The amount of time students can use (and do use) for identifying, exploring, and elaborating new ideas.

Playfulness and Humor. The extent to which the school and the classroom provide a setting for spontaneity and comfort or ease of behavior.

Conflicts (low). The presence (or absence) of personal and emotional tensions in the school or classroom; the presence (or absence) of fighting and aggressive, hostile behavior among students, or overly stern, shouting, angry behavior by the teacher. (This is different from the tensions among ideas that is represented by the "debates" dimension.)

Idea Support. The extent to which new ideas are treated with interest and respect.

Debates. The extent to which encounters and disagreements among viewpoints, ideas, differing experiences, and knowledge will be encouraged in the class.

Risk Taking. The extent to which there is tolerance of uncertainty and ambiguity in the school or classroom.

Classroom Methods: Delivering the Unit

Another important question relating to the use of your individualized unit is, "How should I actually conduct my class to use the unit most effectively?" Once again, the key principle is flexibility. The objectives, learning activities, and evaluation procedures in your unit could be utilized in a number of different ways. There is no reason at all, for example, why it cannot be used as a basis for organizing traditional lecture-discussion classes with individual pupil assignments. You may feel that doing this would defeat some of the purposes of individualizing (and we feel that way, too!), but the point is that the material can be used that way. There may be

Metacognitive Skills for Productive Thinking

Competence

- Knows content or task domain.
- Knows productive thinking processes (tools and underlying concepts and principles; how and when to use tools, individually or in groups).
- Knows and applies the language or vocabulary of productive thinking.
- Attending and focusing; deliberately reviews and analyzes tasks (task demands and constraints; data; new possibilities; criteria and actions).
- Using efficient memory strategies.
- Strong processing (mental pictures, words, emotions; physical sensations).
- Monitors performance and choices; knows how, when to revise or adjust.
- Strong organizing efforts.
- Integrates processes and content in learning and in many life situations.

Confidence

- Believes in effectiveness of processes or methods.
- Believes in self as a facilitator.
- Feels empowered by productive thinking skills.
- Able to change course or redirect without being threatened.
- Comfortable with strategies and language of process.
- Recognizes needs and opportunities for productive thinking.
- Power thinking and affirmations ("I can!").
- Goal-setting and benchmarking.
- Aware of obstacles and ways to avoid or overcome them.
- Predicts and anticipates successful outcomes.
- Views situations as opportunities and challenges.
- Knows own strengths and style and how to make best use of them.

Commitment

- Immerses self in important concerns; passion.
- Takes ownership (responsibility for, and intent to take, action).
- Shows initiative; actively seeks opportunities to apply productive thinking.
- Demonstrates belief in, and regular use of, productive thinking processes.
- Strives for completeness of understanding and continues to work toward that goal.
- Judges results accurately and honestly.
- Engages in "debriefing" and continues to study, learn, and improve.
- Seeks others with whom to learn and share (as colleagues and in mentoring relationships).

Figure 8.1: Metacognitive Skills for Productive Thinking

times when you decide that many of your students share a specific interest and need in some topic, so that you will develop some activities as a large-group instruction approach.

Many of your learning activities probably also lend themselves easily to small-group or team projects, in which a group of four to eight students share the responsibilities for a more extensive project and perhaps share their products with an entire class or grade level.

Of course, you can probably also see from your chart and your learning activities that many of them are well-suited to learning by individual students, using a variety of methods, tools, and techniques. You might develop study guides, for example, each focusing upon a single topic or area in your unit. Or, you might collect several cubes from your matrix together to construct a learning station or learning center. Individual student work, planned and selected from your matrix, can effectively be coordinated by using a contract or learning agreement approach.

Review the Curriculum and Instruction Guideposts in chapter 2. These guideposts will also help you to think about ways you can enhance and vary the ways you deliver or teach the unit.

"But there's just too much ..."

When you have developed an extensive unit plan with a number of topics and an array of specific objectives and varied activities for each topic, you may begin to fear that your plan is just too big, that there are too many possible choices or options, and that it might be confusing or overwhelming for students (or for you). Was it really necessary to do all that planning?

One reason why many teachers limit the instruction in their classroom to routine knowledge and memorization is that they have not planned in advance for other, more complex activities that cannot be accomplished very well "on the spur of the moment." The best way to ensure that your students have opportunities for a richer variety of learning experiences is through careful preplanning. Perhaps the greatest benefit of this planning may be to "open up new possibilities" that you had not thought about previously.

The First Law of Individualized Learning

An important principle underlies much of our modern concern for individualizing instruction. This principle is important enough that we have (only somewhat humorously) called it "The First Law of Individualized Learning." This principle is: A student should not have to sit and watch while other students are learning something that he or she already knows.

The obvious corollary of this "law" is that we must be able to provide opportunities for different students to be doing different things in the classroom at any given time. While these ideas may sound simple, most teachers discover that they can be anything but simple to implement. We believe that planned units for individualized instruction, using the approach you have used in this book, will help you in carrying out these ideas more effectively in your own classroom, even without luxurious "open space" buildings, unlimited aides, or a monumental budget.

Practice Activity on Individualizing Instruction

List some of the unique characteristics of individual students or small groups or clusters of students in your current class or in the class you last taught. For example, one teacher might have noted: (1) Tony loves to read the encyclopedia; (2) Amy learns new material rapidly and easily; and (3) Shawn needs a lot of repetition to master new material, and so on. Make a similar list about several of your students. Then, begin to plan several new ways you could adapt your instructional activities to accommodate some of the characteristics you listed. For example, from the sample list above, you might note:

1. Let Tony take one day a week in reading group time to read the encyclopedia.

2. Create time for Amy to help Shawn with flash cards or a learning game or puzzle.

3. Write some enrichment questions for Amy to work on.

Review the Curriculum and Instructional Guideposts in chapter 2. Try to find ways these guideposts might be useful in thinking about the strengths, talents, interests, and needs of your students.

Then, review the lists of illustrative learning activities in chapter 7 (Figures 7.1, 7.2, and 7.3). Consider how some of those activities might be particularly useful, interesting, and challenging for certain students on your list.

You might identify more than one possibility for some of the students.

Chapter 9: Evaluation & Documentation

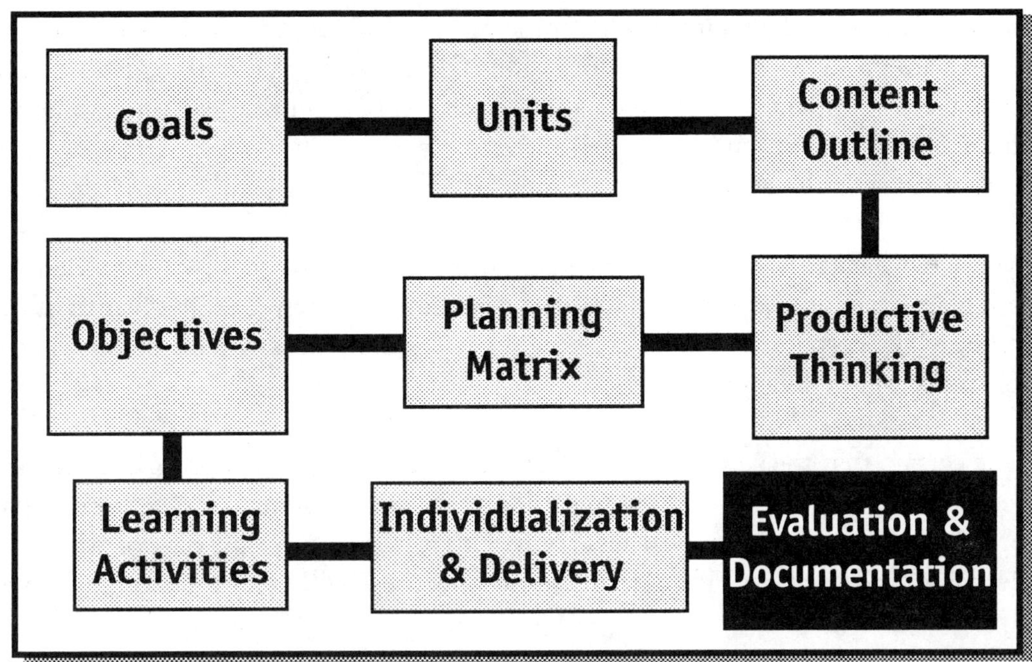

Now we have arrived at "the moment of truth." When any student has been working on one (or several) learning activities, how do you (and how does the student) know when the goal has been reached? How will you decide when more work is needed or when it is time to move on and begin working on some new objective, topic, or even on an entirely different unit of study?

An effective instructional unit should include, for each objective, an evaluation criterion and a statement of how the evaluation will be conducted. This completes the cube that was discussed in chapter 6. The cube for any cell in your matrix includes:

1. one or more objectives for that cell;
2. several learning activities for each objective; and
3. an evaluation criterion or procedure.

Tests may be useful in determining a student's success in completing a certain objective (particularly if it involves Information and the Knowledge Base in the productive thinking dimensions). However, it is not necessarily the case that evaluation always includes, or relies exclusively on, testing. As we will use the term, *evaluation* refers to any appropriate evidence that you can obtain to determine whether the student has learned a certain objective. How do you gather this evidence? How will you decide when a student has met an objective?

Of course, we cannot provide in only a few pages a complete discussion of measurement and evaluation principles. But, we will offer a few guidelines to help you in planning and conducting your evaluation.

1. Check your evaluation procedure carefully for agreement with your objective. It should not only deal with the content of the objective, but it should ask the student to use the appropriate productive thinking dimension. For example, if the objective was intended to engage the student in a complex analysis or comparison of ideas, the evaluation procedure should ask the student to demonstrate his or her ability to analyze, not merely to recall or list specific facts.

2. If you have an expectation about the standards that must be met by the student in completing a project or assignment, state that expectation clearly and in advance of the evaluation. Make certain the students know what the standards will be or on what criteria their work will be judged.

3. Make certain that the standards or criteria you use are fair and appropriate for the objective. Neatness of pupil's work is sometimes misused by teachers, for example, a student's work should be judged for "neatness" only if that is an important part of what he or she is learning, and then only if he or she knows neatness will be evaluated (and how it will be judged).

4. Work with students to define the evaluation criteria and to help them understand what kinds of work will or will not meet the criteria. If you wish to emphasize speed (or completing work within a certain time period), accuracy (doing work carefully and correctly with less emphasis on time), or quality (the inclusion by the students of certain "key" requirements for an acceptable response), you should discuss your expectations and criteria specifically with the students.

5. Many objectives may be written in a way that makes the kind of evaluation that will occur (such as when the student can or cannot perform a certain task) apparent. If you include some objectives of this kind, your evaluation procedure may be nothing more than a checklist to keep a record of work completed by each student. Keep in mind that such a procedure does not take into account quality, accuracy, or appearance unless you specifically include those criteria in advance.

6. Effective criteria for evaluation should be as specific as possible and clearly understood by the student at the beginning of his or her work on the task.

7. Whenever appropriate, students should be encouraged and coached to engage in self-evaluation. Many academic tasks can be checked successfully by students from the later elementary grades upward.

8. Effective evaluation, especially for work that involves the realistic tasks dimension of productive thinking, should provide opportunities for students to demonstrate performance on a task that involves using or applying information. Some performance tasks can be relatively short in duration, depending on the content, so they can be carried out within a single class period or less. Other tasks for more complex objectives may involve extended performance and work over a period of several days, weeks, or even longer.

9. When you are evaluating work in the realistic tasks or the real life opportunities dimensions, include tasks that are authentic or that ask the students to demonstrate their ability to meet the objective through activities that approximate the way that task would be presented and carried out in a real life setting. Be certain to consider the kinds of products and audiences that are appropriate.

10. Evaluation of students' work in the real life opportunities dimension can be an appropriate opportunity for the development and use of portfolios.

Now, let's see if you can apply these guidelines in some examples. Consider the following 10 examples of general objectives. How would you determine specific criteria for evaluating each one? What specific criteria would you use in evaluating these objectives?

1. Home economics students will be able to bake pies.

2. In English class, students will be able to recite poems of their own choosing.

3. Students should demonstrate their ability to add common fractions.

4. Students will name the capitals of the 13 original colonies.

5. Chemistry students will be able to conduct experiments.

6. Students will write essays on the colonies.

7. Students will be able to do push-ups correctly.

8. The first graders will read a full page in their readers.

9. Students will write short essays on assigned topics.

10. Tenth graders will give 10-minute presentations based on library research.

What are the important learning outcomes that the teacher probably wants to emphasize in these objectives? What criteria would be used to determine whether or not students had completed them successfully? Please jot down your own ideas and responses before turning the page.

Here are some responses that illustrate how specific criteria can be used to clarify our purposes and intentions and to make evaluation of student work easier and more efficient and objective. There are other possible responses for all of these, of course. If you thought of other possibilities, share them with other teachers.

1. After instruction on baking, students will produce a pie that (a) does not have a burned crust, (b) comes out of the pan without falling apart, and (c) tastes good according to the teacher or a group of other students. (Criterion here emphasizes Quality.)

2. At the close of the poetry unit, each student will recite a poem of his or her own choice, representing a selected rhyme scheme, an established type of meter, and a central theme or meaning. (Criterion—Quality)

3. After studying addition of common fractions, the student will give the correct sums on at least 8 out of 10 (80%) examples he or she has not previously seen. (Accuracy)

4. The student will be able to make a list of the 13 original colonies and the names of the capital of each one, with not more than two errors. (Accuracy)

5. After studying gases, students will conduct an experiment that verifies Boyle's Law. During the experiment, each student will state a specific hypothesis, collect and record data, reject or confirm the hypothesis, and state one conclusion about the results. (Quality)

6. The student will write an essay analyzing at least three important reasons why the colonists won. For each of the three reasons, the student should provide at least two arguments or references to support his or her choice. (Quality)

7. At the conclusion of the physical fitness unit, each student will be able to do at least 10 push-ups in one minute, touching the mat only with his or her nose. (Speed and Quality)

8. The student will be able to read a page from his or her reader containing at least 100 words within two minutes and make no more than four word identification errors. (Speed and Accuracy)

9. Students will evaluate one another's essays based on criteria they formulated. (Quality).

10. Students will rate each presentation based on evaluation criteria they developed as a group. (Quality)

You might find it valuable to try this exercise with students. Several teachers we know have found it helpful to engage their students in a discussion at the beginning of the school year about what "quality work" means and how it might be recognized. The criteria they formulate together become part of their on-going evaluation dialogue and process throughout the year.

Evaluation Methods

Now that you are familiar with planning and stating evaluation criteria, we should consider the methods you can use to gather and analyze your evidence. We can classify evaluation methods into several broad categories: test responses, checklists and rubrics, rating scales, participation charts, anecdotal records, and portfolios.

Test Responses. The results of tests are most useful in evaluating accuracy of knowledge and comprehension of information or subject matter. Therefore, you are most likely to use this technique to evaluate work in the foundations dimension of productive thinking (especially in relation to information and the knowledge base). Test responses are usually broken down into short-answer responses (multiple choice, true or false, fill-in-the-blanks, matching questions) or essay answers.

The term *formative evaluation* is used to describe testing in which the students use the results to monitor their progress and to determine the work they still need to do. *Summative evaluation* refers to the final evaluation after the completion of all learning activities. It serves as a summary of the progress and learning by the student.

Checklists. Checklists are lists of specific characteristics or behaviors with space for an observer to check whether or not the behavior occurred. These are particularly useful when the specific qualities of a certain outcome can be identified in advance. For example, suppose that you wanted to evaluate a cell in a matrix on a Colonial America unit that deals with the topic of "Food, Clothing, and Housing" and the applying and extending dimension of productive thinking. Suppose that the objective indicates that the student will be able to "apply knowledge about colonial food, clothing, and housing by using it in a new situation or challenge that is provided." A typical learning activity related to this objective might have been for the students to develop an advertisement that might have appeared in *Poor Richard's Almanac* describing a colonial food. A checklist for evaluating the students' performance might have listed the following qualities:

- Is the food clearly described?
- Was the product written in an advertisement format?
- Did the student use color, testimonial, or other advertising techniques?
- Would it be appropriate for *Poor Richard's Almanac*?
- Is it written in a way that will capture the reader's attention?
- Are the prices of the food items included?

A checklist of this type might be given to the students before they complete the learning activities so the criteria do not "come as a surprise" to them when they are evaluated. Part of their work might include discussion of what a good advertisement should include, and their ideas might also be incorporated into criteria that are used in the checklist. Checklists that describe several criteria or standards for acceptable student performance are often described as "rubrics," especially in the current literature on alternative or authentic assessment. A helpful, practical introduction to alternative assessment is *Authentic Assessment: A Handbook for Educators* by D. Hart (1994, Addison Wesley). (*Authentic Assessment of Productive Thinking* [Treffinger & Cross, 1994] deals in greater detail with performance tasks, rubrics, and portfolios.)

Rating Scales. These are lists of specific behaviors that the rater or observer uses to record evaluations of the quality of responses or the degree or extent to which the student's work demonstrates the behavior or meets the criteria. These are also often used as, or described as, rubrics. They require judgment and most often employ a scale that can be quantified (expressed in some numerical form or scale, such as a 1 to 5 scale). For example, consider this objective for the Food, Clothing, and Housing topic in a Colonial America unit: "Design a house that would have been useful in all the colonies." The illustration below shows two rating scale items that might be used to evaluate the students' work for this objective:

1. Did the house design include a way to adjust to the extreme differences in weather that would be found among the colonies?

1	2	3	4	5
Never	Almost Never	Occasionally	Almost Always	Always

2. Rate each feature of the house according to the following scale:

	1	2	3	4	5
a. Windows	Very Poor	Poor	Fair	Satisfactory	Very Good
b. Fireplace	Very Poor	Poor	Fair	Satisfactory	Very Good
c. Room Design	Very Poor	Poor	Fair	Satisfactory	Very Good

Participation Charts. These allow you to count the number of times an individual student contributes in a certain activity, such as a group activity. They are useful when several persons are observed at one time during a group discussion or other activity in which everyone is supposed to contribute or participate. In a Colonial America unit, an objective that calls for students to think of many uses the colonists might have made of corn stalks. The chart on page 69 could be used to keep track of participation by individual students. A tally could be entered for each idea contributed by the members of the group. These could be totaled (a measure of fluency or number of ideas), and other items could also be assessed, such as the number of unusual ideas (a measure of originality in thinking).

Activity: Uses for Corn Stalks			
Name	Tally	Total Number of Ideas	Number of Unusual Ideas
Bill			
Mary			
Chris			
Pat			

Anecdotal Records. These are factual descriptions of a specific behavior, an incident or episode, or an event, written by an observer. Diaries, logs, or journals are similar, but usually refer to records made by the participants themselves, rather than an observer. All are useful for keeping records of unusual events or unexpected happenings that are particularly striking or relevant. For example, if you were interested in whether students were applying what they have been learning about the colonies when they are involved in out-of-class activities, such as a field trip to a museum, you might ask them to keep a record of how many inventions or products they saw in the museum or which art works at a gallery originated in or represented colonial life and times.

Portfolios. A portfolio is a student's unique, personal, and meaningful way of documenting her or his own work and accomplishments within a certain domain or task area for a particular purpose or audience and at a specific time (Treffinger & Cross, 1994). The portfolio can be an important way to document and assess authentic outcomes or accomplishments, particularly for work in the real life opportunities dimension of productive thinking. They are not substitutes for other kinds of evidence or assessment. They are not necessarily the most effective or efficient way to evaluate any and all objectives and activities. And, they are not necessarily only a sampling of a student's "best work." (They might be assembled, for example, to illustrate and document growth or change over time, or to demonstrate breadth or variety in one's area of competence or expertise.) Portfolios might include: products or work samples, testimonials, documentation that verifies participation in specific activities or events, recognitions or awards for one's work, visual documentation (photos, slides, video), audio documentation, scrapbooks, or computer files on disk.

Self- and Peer-Evaluation

Judgments and reports made by students themselves are another valuable source of information about achievement and accomplishments that the teacher might find more difficult (or perhaps not even possible) to evaluate. Such concerns as pupil participation in small-group work, contributions to group projects, acceptance of leadership responsibilities, or evaluating how much and how well one studies or takes notes are areas in which student evaluation can be used.

At first, some teachers who attempt such a process fear that students may be unable to assess their own work or, even worse, that their assessments may be exaggerated or even dishonest. However, when efforts have been made to define objectives carefully and to spell out suitable evaluation criteria in advance, most students will be able to engage in self-evaluation effectively. In fact, there may be a tendency for the students to be more demanding on themselves than the teacher might have been.

Before requiring students to evaluate either their own work or their peers' work, the teacher should verify that students have learned some basic skills and experience in making evaluations. Students should be able to demonstrate that they:

- can read test or assignment directions and listen to oral directions;
- can interpret test items carefully so they understand what is required for a correct response before they make a judgment;
- have skill in detecting and responding to the most obvious or reasonable interpretation of a test item, rather than to an interpretation that, while possible, was probably unintended;
- are able to select a reasonable answer to each test item, even when they may know very little about item content;
- have experience in recording their observations, as would be required in compiling anecdotes, keeping diaries, or filling out checklists and rating scales;
- are aware of the "pitfalls" and biases that can affect the use of rating scales, checklists, or other techniques that involve a degree of subjective judgment, (for example, one of the most difficult tasks in self- or peer-evaluation is rating each item separately and independently, rather than on the basis of an overall impression or feeling—the "halo" effect);
- have no more than a moderate amount of "test anxiety" or feelings of inadequacy and helplessness in situations requiring performance to be demonstrated to others (Help your students by modeling appropriate behavior in evaluations and guiding students in attending specifically to the evaluation task.); and
- understand the criteria to be employed and the methods that are to be used to apply those criteria to the products or performances being evaluated.

Any of the evaluation methods described in this chapter can be used by students, either to evaluate others or to make fair judgments about themselves. If teachers indicate clearly how the evaluation is to be conducted and how the information will be used, if they guide students in learning and using evaluation methods and techniques, if they use formative evaluation as an aid to student learning, and if they ask for judgments that are within the students' level of competence, then asking students to make assessments of their own or others' performance can be a valuable and appropriate part of the evaluation process.

Now, you should be ready to plan some specific evaluation criteria and procedures for the unit you are planning. Your matrix has been growing steadily since you began with a simple content outline in chapter 3. By now, you should have a complete set of written objectives and a variety of learning activities incorporating many different dimensions of productive thinking. Your final task is to take the cells in your plan and ask: "How will I evaluate whether the student has completed this objective successfully? What kinds of evidence would be most useful? What criteria should be used? What method(s) would be most appropriate to use to obtain the evidence or documentation?"

When you have completed this task for your matrix, you will have a thoroughly planned instructional unit that will lend itself readily to individualized learning for your students.

References

Costa, A. L. (Ed.). *Developing minds: Programs for teaching thinking.* (Rev. ed.). Alexandria, VA: Association for Supervision and Curriculum Development

Dunn, R., & Dunn, K. (1978). *Teaching students through their individual learning styles.* Reston, VA: Reston Publishing.

Dunn, R., Dunn, K., & Treffinger, D. J. (1992). *Bringing out the giftedness in your child.* New York: John Wiley.

Feldhusen, J. F., & Treffinger, D. J. (1985). *Creative thinking and problem solving in gifted education* (3rd ed.). Dubuque IA: Kendall-Hunt.

Isaksen, S. G., Dorval, K. B., & Treffinger, D. J. (2000). *Creative approaches to problem solving* (2nd ed.). Dubuque, IA: Kendall-Hunt.

Isaksen, S. G., Treffinger, D. J., & Dorval, K. B. (1996). *Climate for creativity and innovation: Educational implications.* Sarasota, FL: Center for Creative Learning.

Renzulli, J. S. (1994). *Schools for talent development: A practical plan for total school improvement.* Mansfield Center, CT: Creative Learning Press.

Stanley, J. S. (1980). On educating the gifted. *Educational researcher, 9*, 8–12.

Treffinger, D. J., & Cross, J. A. (1994). *Authentic assessment of productive thinking: Training module* (Field Test Ed.). Sarasota, FL: Center for Creative Learning.

Treffinger, D. J., Cross, Jr., J. A., Feldhusen, J. F., Isaksen, S. G., Remle, R. C. & Sortore, M. R. (1993). *Productive thinking handbook: Volume I— Rationale, Criteria and Reviews.* Sarasota, FL: Center for Creative Learning.

Treffinger, D. J., & Feldhusen, J. F. (1996). Talent recognition and development: Successor to gifted education. *Journal for the Education of the Gifted, 19*(3), 181–193.

Treffinger, D. J., Feldhusen, J. F., & Isaksen, S. G. (1990). Organization and structure of productive thinking. *Creative Learning Today, 4*(2), 6–8.

Treffinger, D. J., Feldhusen, J. F., & Isaksen, S. G. (1996). *Guidelines for selecting or developing material to teach productive thinking.* Sarasota, FL: Center for Creative Learning.

Treffinger, D. J., Isaksen, S. G., & Dorval, K. B. (2000). *Creative problem solving: An introduction* (3rd ed.). Waco, TX: Prufrock Press.

Treffinger, D. J., & Nassab, C. A. (2000). *Thinking tools lessons.* Waco, TX: Prufrock Press.